Make:
EASY
1+2+3
PROJECTS

Make:
EASY
1+2+3
PROJECTS

The Editors of Make:

MAKER MEDIA
SAN FRANCISCO, CA

From the Pages of Make:
Easy 1+2+3 Projects
By the Editors of Make:

Printed in Canada.

Published by Maker Media, Inc.,
1160 Battery Street East, Suite 125,
San Francisco, California 94111

Maker Media books may be purchased for educational, business, or sales promotional use. Online editions are also available for most titles (http://safaribooksonline.com). For more information, contact our corporate/institutional sales department: 800-998-9938 or corporate@oreilly.com.

Publisher: Brian Jepson
Editor: Roger Stewart
Production Editor: Happenstance Type-O-Rama
Proofreader: Happenstance Type-O-Rama
Interior Production: Cody Gates & Kate Kaminski, Happenstance Type-O-Rama
Cover Designer: Brian Jepson

November 2015: First Edition

Revision History for the First Edition
2015-11-15: First Release

See http://oreilly.com/catalog/errata.csp?isbn=9781457186899 for release details. Make:, Maker Shed, and Maker Faire are registered trademarks of Maker Media, Inc. The Maker Media logo is a trademark of Maker Media, Inc.

978-1-68045-044-6
[TCP]

CONTENTS

Part III: Science and Electronics **69**

Part IV: Home and Outdoors 119

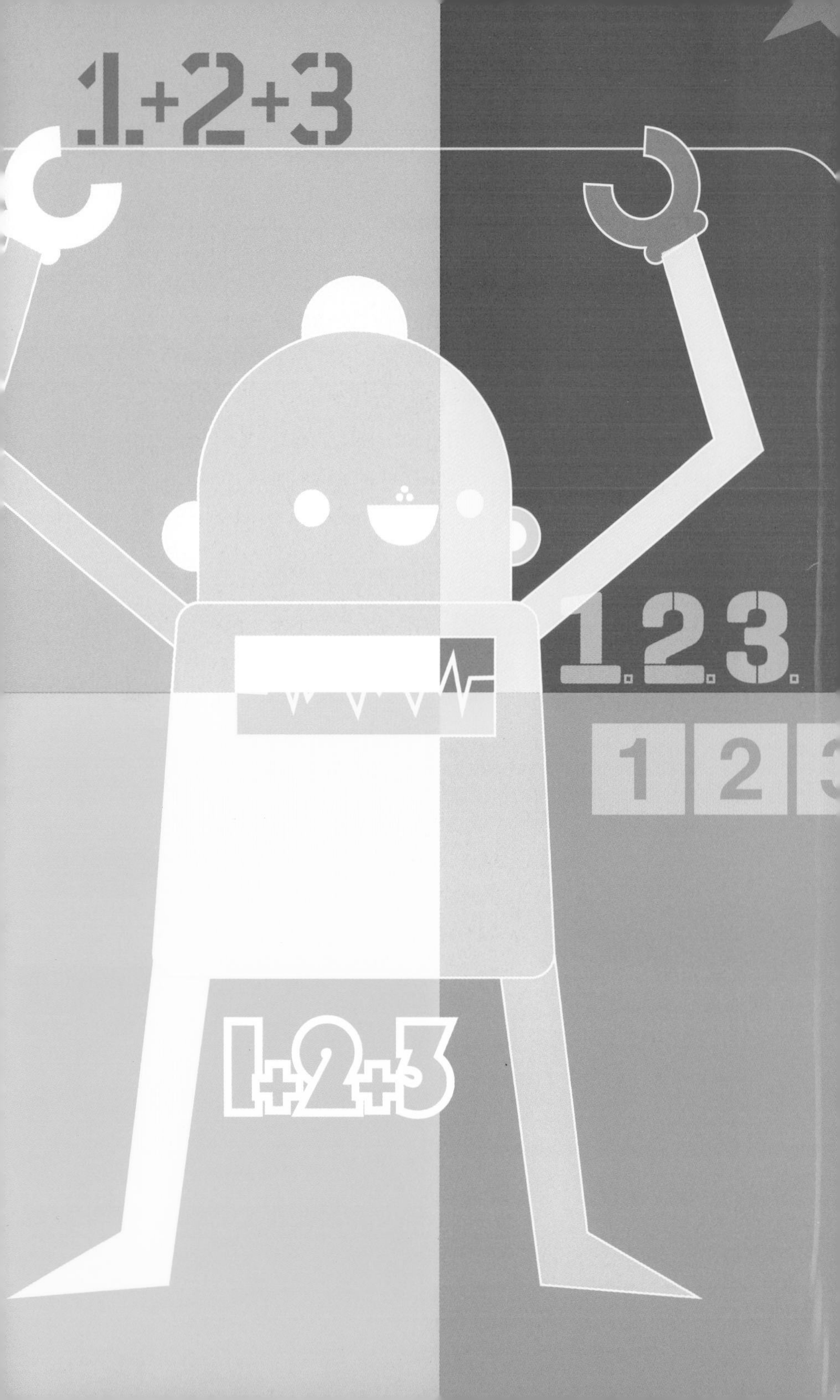
1.+2+3
1.2.3.
1 2 3
1+2+3

Preface

"1+2+3" has been a regular feature of *Make:* magazine since its earliest days.

As the name implies, the projects are all "as easy as 1, 2, 3." Each project is explained with a handful of easy-to-follow, step-by-step instructions and simple, clear illustrations.

In the pages that follow, you'll find projects that are fun, artistic, useful, and educational. Most are perfect rainy-day activities that you can do with kids. Parents and educators will find lots of suitable home and classroom activities. A handful of the projects may call for garage workshop tools such as a saw or a drill, but many require nothing more than scissors and tape.

It's easy to follow along and re-create the projects just as the authors have presented them. But, of course, you can also use them as springboards for your own ideas, mods, and hacks.

Let your creativity flow!

Part I: Toys and Games

YOUNG IMAGINATIONS get their greatest workouts from toys and games that they make themselves or with a little help from parents and educators. The simple but sturdy design of Ed Lewis's "Fast Toy Wood Car" is a great example. You can make it with a few basic workshop tools and inexpensive materials; and it's a perfect platform for putting imaginations to work on *new* designs and modifications.

Paul Rawlinson explains how to make a working miniature canon using party poppers. Gus Dassios provides instructions for a different kind of popper—one that will liven up your next board game. A design engineer by profession, Gus says, "When I'm not sleeping (and even *then* sometimes), I'm always thinking of the next thing I want to make."

Edwin Wise is a master of Halloween props and scares. His startle-inducing "Boom Stick" is reprinted in *The Best of Make: Volume 2*. In "Ping-Ponger," Edwin shows you how to make something a little less scary but still full of potential surprises.

Julie Finn says her children are the inspiration behind most of her Maker projects. Julie's "Custom Memory Game" is a great way to challenge young minds. Check out her blog posts online at *Crafting a Green World* for more great ideas for kids.

Ben Wendt, a lifelong tinkerer, has also been inspired by his kids to take his Maker projects on further explorations. You may be inspired by his "Remote-Controlled Camera Mount" project to make a "Hale's Tour" of your own neighborhood.

Ian Gonsher is an artist, designer, and educator on the faculty in the school of engineering at Brown University. His "Coffee Shop Construction Toy" demonstrates how to enlist simple everyday objects as design tools. George W. Hart is research professor in the engineering school at Stony Brook University, as well as a freelance mathematical sculptor and designer. His "Two-Cent Wobbler" (aka two-circle roller) is addictively fun to watch roll down an incline, but for the mathematically minded it also demonstrates some important principles.

Along with his his wife, Hazuki Kataoka, David Battino writes and performs Japanese storytelling books for children. Watching sword fighters at a Japanese cultural festival inspired him to create the "Safe Bamboo Swords" project.

Brian McNamara ("Alien Projector") makes small synthesizers, loopers, and various noise- and image-makers. An article in *Make:* encouraged him to start selling his creations on Etsy. A *Make:* editor subsequently saw his R-Tronic sequencer on Etsy and asked him to write about it for the magazine. Thus the circle was complete!

These and many more DIY projects are waiting for you in the pages that follow. It's time to get started making . . . and having fun! ◢

Fast Toy Wood Car

By Ed Lewis ■ Illustrations by Julie West

YOU WILL NEED

Plywood ¼", 11"×14" or more
Bolts ⁵/₁₆, 4" (2)
Locknuts ⁵/₁₆" (2)
Spacers ½" (4)
Inline skate wheels with bearings (4)
Wood glue
Laser cutter or jigsaw, router, or coping saw, and drill with ⁵/₁₆" bit
Clamps
Cutting templates

LOTS OF MY FRIENDS HAVE KIDS, and that means lots of birthdays. I wanted to have a custom present that's easy to make and has lots of room to play, in terms of design. A toy car fits perfectly. So I can build cars and make kids happy? Win-win!

1. Cut the plywood.

Download the templates from makezine.com/projects/make-32/fast-toy-wood-car/ and use them to cut the plywood body. Use a laser cutter, or cut with hand or power tools. Sand edges.

2. Assemble.

Run each bolt through a wheel, a spacer, the car body, another spacer, another wheel, and a locknut to cap it off. (Remove the wheel's internal spacer if necessary.) The car is ready to roll! If you want, change a layer or two, or even redesign the whole thing.

3. Glue.

Take the car body apart and apply wood glue between the layers. Reassemble, clamp, and let dry. You now have a toy car that's ready for tons of abuse. It can go very, very fast. Little kids will have no problem moving it around, and bigger kids will enjoy whipping it off ramps to see how it performs.

Going Further

There's lots of room for customization. Make the body profile realistic or more abstract. Give the car a front and back, or make it symmetrical. Play with the wheel size and the distance between the front and back wheels. Stain or paint can liven your car up, as well as extra details such as names or stickers. Make your car what you want it to be! ◪

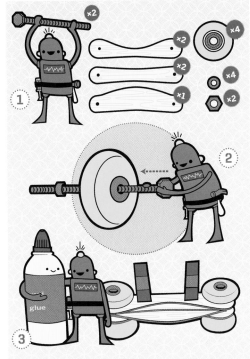

WHEE!

Ed Lewis lives in Oakland, CA, with his wife, two sons, two cats, and a shed full of tools.

Action Root Beer Pong

By Cy Tymony ■ Illustrations by Damien Scogin

YOU WILL NEED

Styrofoam or paper cups

Ping-pong balls

Toy motor and paper clip, or micro-vibration motor
  RadioShack #273-107, radioshack.com

Paper clips, jumbo (3)

Button cell battery CR 2032, but any 3V type will do

Copper wire

Tape

Pliers

PLAYING THE POPULAR BEER PONG GAME can be enjoyable for a while, but as your skill improves, the challenge and fun diminish. After some practice, tossing a ping-pong ball at a cup just isn't too difficult. Bring back the fun by adding motion, with your own pong toy that scuttles about.

1. Make the legs.

Bend 2 paper clips into curved "C" shapes, with legs, and tape them to the sides of a cup.

2. Connect the motor.

Using a small motor removed from a toy, bend and press a paper clip around the gear shaft to make it off-balance. (Alternatively, you can use a ready-made mini vibrating motor found in an old pager or cellphone.) Test the motor with a 3V button cell battery by pressing its wires onto both terminals.

Tape the motor and battery to the bottom of the cup. Tape a short length of copper wire to one side of the battery. Tape one wire connector from the motor to the other side of the battery. Tape the 2 loose wires (from the motor and the battery) to the side of the cup so they can be twisted together, acting as a switch.

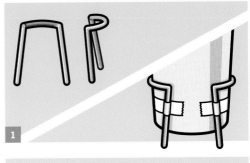

3. Play ball!

Turn the cup right side up, twist the 2 switch wires together, and the cup should vibrate and move on a flat surface. You can bend the cup's paper-clip feet so it will move in a desired pattern. Now try to toss balls into the moving cup from various distances.

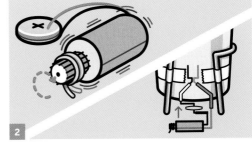

Going Further

To add to the challenge, use a paper clip to attach another cup (or two) to the side of the Action Root Beer Pong cup. Draw score numbers on them to make it more competitive. ◪

Cy Tymony (sneakyuses.com) is a Los Angeles–based writer and is the author of *Sneakier Uses for Everyday Things*.

Mini Foosball Game

By Cy Tymony ▪ Illustrations by Damien Scogin

YOU WILL NEED
Microwave popcorn box 3-bag size
Straws (4)
Jumbo paper clips (8)
Gumball or jawbreaker candy
Scissors
Tape

FOOSBALL (TABLE FOOTBALL) GAMES are a lot of fun, but they're usually found at bars or clubs. That shouldn't stop a true Maker from enjoying foosball at home.

With just a few straws, paper clips, and a common 3-bag microwave popcorn box, you can quickly put together your own Sneaky Mini Foosball Game.

1. Prepare the cardboard.

Open and flatten the popcorn box. Cut off the front of the box, but don't trash it—it'll be used to make sneaky game paddles. Fold the cardboard back into a box shape and tape the ends together. Be sure to cut holes at each end for the game ball to exit.

2. Prepare paddle guides and paddles.

Affix 8 paper clips onto the long sides of the box, 4 per side, evenly spaced. These will act as paddle guides.

Using the cardboard front you saved earlier, cut 6 paddles approximately ¾" wide by 2" long, oval at one end and square at the other.

3. Attach paddles and straws.

For the "players," tape 2 paddles near the center of 2 straws as shown. For the "goalies," tape 1 paddle each at the center of 2 straws. Slip the straws into the paper clip holders with the goalies at each end, and you're done. (You can also slide the straws through the guides first and then attach the paddles—either way works.) Place the candy ball on the game board area and start the action. ▨

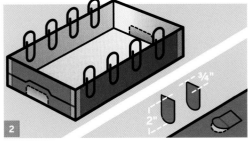

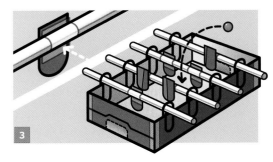

Going Further

To make a Sneaky *Micro* Foosball Game, go to http://makezine.com/projects/make-28/micro-foosball-game/.

Cy Tymony (sneakyuses.com) is a Los Angeles-based writer and is the author of *Sneakier Uses for Everyday Things*.

Party Popper Toy Cannon

By Paul Rawlinson ■ Illustrations by Julie West

YOU WILL NEED

Wood block **77mm long × 43mm wide, with the top chopped at a 15° angle**
Wood circles **27mm diameter × 5mm thick (4)**
Wood dowels **6mm diameter × 300mm long (3)**
Copper pipe **15mm diameter × 110mm long**
Pipe fitting **compression cap, brass, 15mm**
Copper pipe clips **15mm (2) aka saddle clips**
Wood screws **small (4)**
Cotton balls
"Party popper" confetti shooters **(5)**
Bottle lid **plastic, less than 15mm diameter**
Wood glue
Drill
Drill bit **hole saw, 27mm**
Drill bit **wood, 7mm**
Drill bit **metal, 2mm**
Adjustable wrenches **(2)**

MOST LITTLE BOYS PLAY WAR, BUT YOU CAN'T GET YOUR HANDS ON THE REALLY COOL TOYS UNTIL YOU'RE AN ADULT. I wanted

> **⚡warning:** Take all necessary precautions and wear safety equipment during this build. Handle party poppers carefully, as they do contain explosives, and never fire at anyone.

to bridge that gap a little by making a working toy cannon. This was a quick model made with very little planning, but by taking your time on the details, you could easily make a better version.

1. **Drill.** For the wheels, drill 4 cores (27mm) from 5mm wood using the hole saw, then drill 7mm holes in the center of each. » Drill two 7mm holes through the width of the wood block, in the bottom front and rear, to accept axles. » Drill a 2mm hole through the compression cap.

2. **Assemble the cannon.** Insert the axle dowels and use a blob of wood glue to attach each wheel permanently. » Attach the copper pipe to the top of the block using the saddle clips and screws, leaving a slight overhang on both sides of the cart.

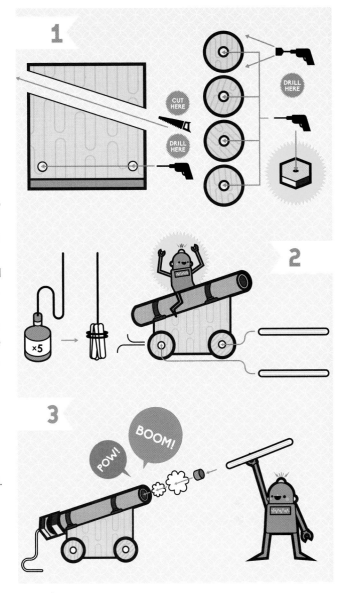

3. Load and fire!

Disassemble 5 party poppers and tie together the 5 mini explosives. Feed one string through the hole in the compression cap.

» Using the pair of adjustable wrenches, tighten on the compression cap so that the "olive" ring binds, then loosen it slightly so that later you can remove and retighten it by hand.

» Using the third dowel, ram a cotton ball or two down the barrel of the cannon, so they sit over the explosives. Lastly, add the ammo (I used a lid from a soy sauce bottle) and you're ready! Pull the string to fire.

You can easily make your own adjustments to this cannon design to make all different types and styles. ▨

Paul Rawlinson makes random projects and shares the results on his site, go-repairs. blogspot.co.uk and YouTube channel, youtube.com/user/gorepairs/.

Dice Popper

By Gus Dassios ▪ Illustrations by Julie West

THIS IS MY VERSION OF A CLASSIC WAY to "roll" the die for a board game. The key to making this special die agitator is the clear plastic capsule. I found the right size, 2" diameter and 1$^7/_8$" tall, in a vending machine—it's the kind that holds inexpensive toys or trinkets.

1. Cut the wood.
Parts A, C, and E are cut of ¼"-thick wood, while parts B and D are ¾" thick. Cut all 5 pieces to 3½"×3½".

2. Drill the holes.
Holes have to be drilled in all pieces except part E. » For part A, cut a large hole in the center, anywhere from 1$^7/_8$" to 2" in diameter, depending on the size of the domed half of your capsule. Then drill 4 countersink holes about ½" in from each corner, for the wood screws. » For parts B and C, cut a 2$^1/_8$"-diameter hole in each center. These holes will house the bottom part of the capsule. For part D, drill a $^3/_8$"-diameter hole in the center, which provides a guide for the spring.

3. Assemble.

Nail together parts B through E with the finishing nails, avoiding the corners where the screws will go. After this bottom block is assembled, insert the spring. Place a die into the capsule, snap it, and lower it into the wood block. Part A will cap everything. It requires a slight push to keep it in place as you screw in the wood screws. » After that, you're ready to go. Good luck! ◪

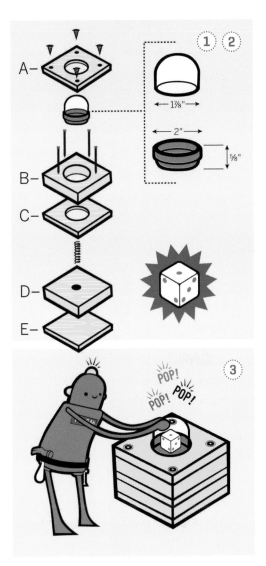

Gus Dassios lives, designs, and builds in Toronto, Ontario.

Ping-Ponger

By Edwin Wise

THE PING-PONGER uses almost half of a rubber racquetball as a disc spring that's bistable (it can be at rest in 2 possible states) to propel a ping-pong ball from a compact PVC launcher.

1. Make the PVC parts.

Cut all PVC pipe pieces to length. For the body, cut the ends off the snap tee so it's 2½" long, then press-fit the female adapter into the bottom of the tee and cut it off flush.

Glue the backstop rings together so that one edge is flush.

Bevel the inside edges of the barrel pieces with a file, so they curve to match the shape of the racquetball. Glue the 2 barrel pieces together.

For the handle, sand the 1" repair coupling so it fits into the 1½" pipe and glue it in place, then glue the tip of the 1" pipe into this coupling.

2. Make a disc spring.

Seat a racquetball into the beveled end of the barrel. Trace a line around the ball (parallel to the seam) and cut on this line, leaving a dome. This is your disc spring.

3. Put it all together.

Glue the backstop into the body (the snap tee), flush with the back end.

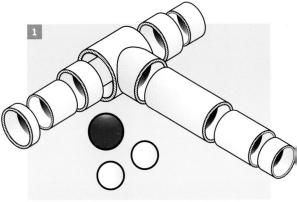

Set the rubber spring on the backstop inside the body. Slip the barrel in place over the spring, leaving just enough room for it to spring forward and back. Don't glue the barrel; you can remove it to change the spring.

Slip (don't glue) the handle into the bottom of the body. The handle also acts as a ping-pong ball holder and ramrod.

Now Ping-Pong Away!

Photograph by Ed Troxell

Put a strong, name-brand ping-pong ball into the barrel, and push it into the spring using the handle, until the spring sticks open (back), gripping the ball. Through the backstop, poke the tensed back of the spring to make it un-spring. Pong! ◪

Edwin Wise is a software engineer and rogue technologist with more than 25 years of professional experience, developing software during the day and exploring the edges of mad science at night. simreal.com

Building Block Picture Puzzles

By Jason Poel Smith ■ Illustrations by Julie West

YOU WILL NEED
Building blocks interlocking
Knife
Photo paper self-adhesive or you can use regular photo paper and spray adhesive
Computer and printer

TWO OF MY FAVORITE TOYS ARE BUILDING BLOCKS and picture puzzles. So naturally I decided to combine them to make building block picture puzzles.

1. Build the block frame.
Start by attaching a bunch of blocks together to form a surface for the picture. This can be a simple rectangle or any fiendishly complicated shape that you want to use. » Press the blocks firmly together so that the picture can be applied evenly.

2. Apply the picture.
Next, select a picture to use for the puzzle. » Scale it to the appropriate size on your computer, and then print it out on self-adhesive photo paper. Alternatively, you can use regular photo paper and spray adhesive. » Carefully apply the picture to the block frame. Press the picture down firmly to ensure that it's well adhered to the blocks.

3. Cut out the pieces.
Finally you'll need to cut out the individual blocks. » Take a sharp knife and gently fit it into the groove between 2 blocks. Carefully slide the blade along the groove to cut the picture into pieces. » Once all the blocks are cut out, press down on all the cut edges to make sure they're properly adhered to the block.

You can use this basic procedure to make two-sided puzzles or even complex three-dimensional picture puzzles. Try it out and have fun. ◪

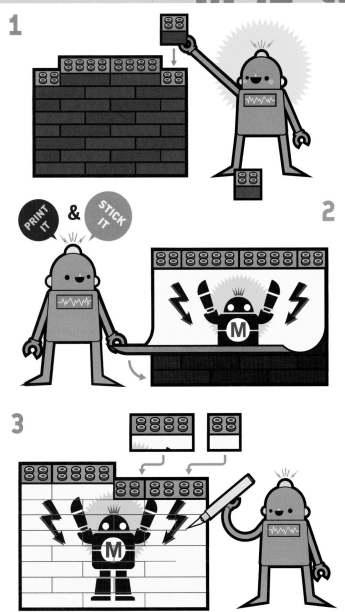

Jason Poel Smith makes the "DIY Hacks and How Tos" video series on *Make:*. He is a lifelong student of all forms of making, and his projects range from electronics to crafts and everything in between.

Custom Memory Game
By Julie A. Finn

YOU WILL NEED

Paper plain, colored, or printed, for making matching image pairs
Scissors
Glue stick
Laminating sheets or a laminating machine
Fancy paper such as scrapbooking or wrapping paper, for the faceup and facedown sides of your memory cards

MEMORY IS BASICALLY THE PERFECT GAME. A kid can play it solo or in a group, it can be easy or challenging, and it's educational to boot. Give an old standby some new flavor and create a perfect rainy-day activity when you make your own memory game out of materials you already own.

1. Prepare your pairs.

Cut out several sets of matching image pairs from paper, such as numbers, colors, or words. (Here I used paint swatches.) If necessary, trim them all to be approximately the same size or to cut off extraneous words or images.

2. Prepare your template and backing.

Find a template for your memory cards—cassette tape, coffee mug, etc.—and cut out identical faceup sides from your fancy paper for each card in your game. Also cut out a facedown side for each card—this can be from the same paper as your faceup side, or different. They don't have to be identical, because they won't be facing up at the start of the game.

3. Glue and laminate.

With a glue stick, lightly stick your faceup and facedown sides together (facing up and down!), then stick one memory game image on each facedown side.

Laminate your cards for durability, or take them to a copy shop to be laminated.

Use It.

While simple identical color matches work well for a memory game, consider tweaking your game toward a specific educational experience or a specific kid. Try matching colors with color words, for instance, or pictures with Spanish words, or math equations. Keep your kid guessing! ◢

Photography by Julie A. Finn

Julie A. Finn blogs about all the wacky hijinks involved in parenting and the crafty life at craftknife.blogspot.com.

Candy Alert
By Cy Tymony ■ Illustrations by Tim Lillis

YOU WILL NEED
Gummy candy with light-up tongs
R/C car transmitter
Noisemaker such as one from a candy toy cellphone

CANDY MAKERS ARE INCLUDING INNOVATIVE EXTRAS with their products that sneaky scavengers can reuse for projects. A cursory look around the candy aisle reveals spring-loaded containers, light- and sound-producing cellphone toys, battery-powered fans with amazing light shows, even tongs that light up when you squeeze them to grab gummy candy. The batteries, switches, LEDs, and motors included in just these 4 packages would cost about $10 if purchased separately. Here's how to easily modify some of these useful parts for sneaky projects, in this case an intruder alert.

1. Tape the toy tongs' switch to a door.
The light-up tongs include a watch battery, an LED, and a pressure switch that activates when you squeeze the tongs. This switch can be removed to act as a security trigger when it's positioned with tape to a door, window, cabinet, or drawer (Figure 1).

2. Wire the switch to an R/C car transmitter.
Connect the tongs switch's 2 wires to an R/C car's transmitter activator button (see "Sensor SenseAbility," page 90) so that it can alert you when doors or windows have been breached (Figure 2).

3. Wire the R/C receiver to a noisy alarm.
Now connect the R/C receiver's output contacts to a noisemaker—such as the candy toy cellphone. You can also wire the tongs switch directly to the toy cellphone (Figure 3).

Going Further

A candy fan toy can be converted into a motorized car. A spring-loaded candy stick makes a great sneaky security device, triggering the toy cellphone alarm like a Rube Goldberg contraption.

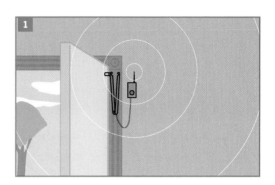

Magazines sometimes include high-tech inserts to promote products, and these too have parts ripe for the reusing. NBC once placed Bionic Woman TV show promotional inserts in major magazines. These inserts included 2 watch batteries on a printed circuit board, connecting wire, and a slide switch that lit a

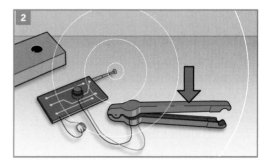

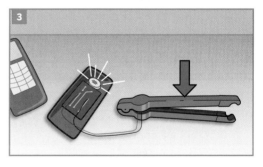

super-bright LED when you turned the page. If purchased separately these parts would cost nearly $10, and you can easily put them to use as alarms, educational quiz testers, and more. ☑

Cy Tymony (sneakyuses.com) is a Los Angeles-based writer and is the author of *Sneakier Uses for Everyday Things*.

Coffee Shop Construction Toy

By Ian Gonsher ■ Illustrations by Julie West

YOU WILL NEED
Coffee stirring sticks (12)
Straws (3)
Scissors optional

COFFEE SHOPS ARE GREAT PLACES TO EXPLORE your creativity and come up with new ideas. A little caffeine, some time to play, and the materials at hand can go a long way. With a bit of imagination you can use this construction method to build many other kinds of forms, and this basic component can be combined with other elements to create larger and more impressive pieces. The baristas don't tend to mind, especially if you tip generously.

1. Collect and cut.

After you've customized your caffeinated beverage of choice at the condiment bar, pick up a few extra items: 12 coffee stirrers and 3 straws. Cut the straws in half, creating 6 connector pieces (if you don't have scissors, you can do this with 6 full-length straws as well).

2. Build an element.

Build a basic triangular element with 3 coffee stirrers and 3 straw pieces. Push 2 coffee stirrers, pressed flat against each other, into the end of each of the straws to connect 3 coffee stirrers together.

TIP: Soak the stirring sticks in water if you have trouble getting them to bend.

3. Combine the elements.

Build and attach 3 more triangular elements together until you get a larger triangle. Fold the structure upward, and connect it with the final 3 stirrers. This should give you a cube-like structure. ◪

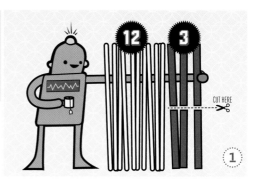

Ian Gonsher is an artist, designer, and educator on the faculty in the School of Engineering at Brown University. His teaching and research focus on creative practices across disciplines and the development of pedagogical strategies for teaching creative thinking. iangonsher.com

Alien Projector

By Brian McNamara

YOU WILL NEED

LED green
Resistor 330R
Switch SPST panel mount
6" of wire
9V battery and battery clip
1¾" PVC pipe, 6½" long
³⁄₁₆" balsa wood sheet, 6"×4"
³⁄₈"×½" balsa wood stick,
    4" long
Heat-shrink tubing optional

Saw
Wire cutters
Wire strippers
Sharp knife
Hot glue gun
Solder
Soldering iron
Download templates from:
    cdn.makezine.com/make/
    16/123_alien_template.jpg

THIS SIMPLE PROJECTOR SHINES an image of an alien on the wall. It uses an LED as the light source and projects an image varying in size from a few inches to several feet. The simple circuit consists of only a battery, resistor, switch, and LED.

1. Build the basic circuit.

Trim the resistor and LED leads to ¼" long, and solder 1 end of the resistor onto the negative lead of the LED. Cut about 6" of wire, and solder it to the positive lead of the LED.

Put a piece of heat-shrink tubing over the black (negative) wire of the 9V battery clip, then solder the wire to the other leg of the resistor. Solder the red (positive) wire from the battery clip to the middle terminal of the switch. Place heat-shrink tubing over this wire if you wish. Solder the positive LED wire to one of the outside pins of the switch.

2. Build the case parts.

Print the templates of the front and back plates, cut them out, and tape them to the balsa wood sheet. With a sharp knife, cut out the inside of the template first, then the outside. Use the cross on the back template to mark the hole for the switch.

Cut two 1¾" lengths of balsa wood stick. With the knife, drill a ³/₁₆" hole in the middle of one, and a ¼" hole in the middle of the other.

Drill a ¼" hole in the middle of the balsa back plate where it's marked, then glue the stick with the ¼" hole to the back plate.

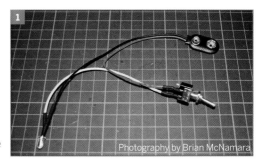

Photography by Brian McNamara

3. Put it all together.
Glue the LED into the stick with the 6" hole. Fit the switch to the back plate. Fit the LED stick into the middle of the PVC pipe, then glue the front plate onto the pipe. Snap the battery into the battery clip, then fit the back plate into place.

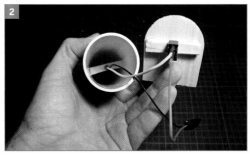

To set up the projector, place it in a dark room 1'–4' from the wall. Move

it closer to the wall for a smaller image, and farther away for a bigger image. The best thing about this project is that you can cut out any simple image and project it onto the wall; perhaps pumpkins for Halloween and a tree for Christmas. ◪

Brian McNamara lives near Canberra, Australia. By day he works in a university workshop designing and repairing equipment for a biological research facility; by night he designs, hacks, and bends kids' toys and musical instruments.

Safe Bamboo Swords
By David Battino ■ Illustrations by Julian Honoré/p4rse.com

YOU WILL NEED
Bamboo poles about 1" in diameter by 6' long
Foam pipe insulation The kind with a slit along one side is best.
Duct tape
X-Acto knife to trim foam
Saw (optional) to cut bamboo

AT AN OUTDOOR FESTIVAL, MY SONS AND I saw a group of teens play-fighting with padded swords. That looked like a lot of fun, so we tried making our own. It was remarkably easy and inexpensive—each sword costs about $4.

1. Cut the bamboo (if desired).
Six-foot swords are great, but we also made two 3-footers.

2. Wrap it in foam.
Squeeze the foam around the pole, leaving about 12" uncovered for the handle. Allow the foam to extend 1"–2" beyond the other end of the pole so the tip is padded.

3. Add duct tape for strength.
The foam we chose has adhesive along the slit, but a few wraps of duct tape every foot or so will ensure that it doesn't split open. Some people wrap the whole thing in duct tape. Covering the handle with duct tape is also an option.

David Battino makes music (batmosphere.com) and Japanese storycards (storycardtheater.com).

Use It.

We battle using rules based on Monty Python's Black Knight: a chop to your opponent's arm or leg disables it; keep fighting until one of you has no limbs left. In the photo below, Sir Christopher has lost both legs and King Toma has lost an arm.

It's great, silly fun to fight while hopping on one leg or gripping the sword between your knees because you've lost both arms. Plus, the low mass of the poles and the thick padding make sparring fairly safe.

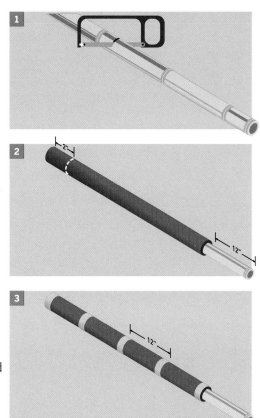

Photograph by David Battino

Boomerang

By Cy Tymony ■ Illustrations by Dustin Amery Hostetler

YOU WILL NEED

Corrugated cardboard
Foam rubber
Tape
Scissors

HAVE YOU EVER WONDERED HOW BIRDS FLY or how sailboats can sail into the wind? In "Origami Flying Disc," I show how to make a flying disc using the Bernoulli principle to generate lift; here we use the same principle to create a boomerang out of ordinary stuff like cardboard.

1. Cut the cardboard to the size and shape shown. Each wing of the boomerang should be 9"×2".

2. Cut 2 foam pieces into oval shapes about 6"×2" with 1 side rising, as shown in Figure 2. The rising shape should resemble the side view of an airplane wing. Place the oval foam pieces on the leading edges of the boomerang and secure them with tape. The foam creates a curved shape on the boomerang wing, causing air to move faster across its top area than the bottom surface. This will produce lift for the boomerang.

NOTE: Look carefully at the placement of the ovals on the wings in Figure 3 before taping them.

3. Throw the boomerang like you were going to toss a baseball. Throw it straight overhead, not to the side. The boomerang should fly straight and return to the left. Experiment with different throwing angles to obtain a desired return pattern. If the boomerang does not return properly, add extra weight by using thicker cardboard or by taping coins, evenly spaced, to its surface. You can also experiment with thicker pieces of foam to improve lift. ◩

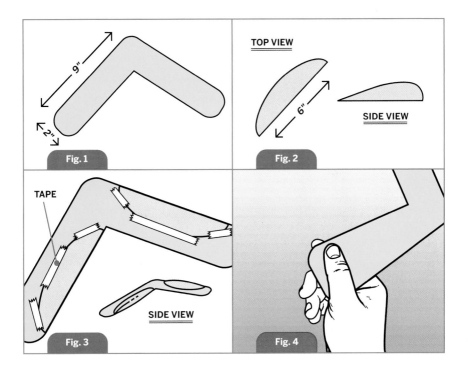

Cy Tymony is the author of the *Sneaky Uses for Everyday Things* book series. sneakyuses.com

Paper Water Bomber

By Ewan Spence ■ Illustrations by Gerry Arrington

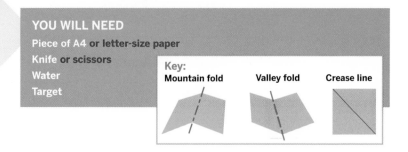

YOU WILL NEED
Piece of A4 or letter-size paper
Knife or scissors
Water
Target

Key:

Mountain fold	Valley fold	Crease line

THIS WINGED ORIGAMI MISSILE WITH FRONT-LOAD TANK DELIVERS A WET PAYLOAD.

1. Make the body.

a. Cut the paper to make a sheet with a length/width ratio of about 2:1.

b. At one end, fold each corner down to the opposite edge, then unfold, to make a square X crease. Turn over and fold the top edge down to bisect the X. Turn over again, push the edges in, and flatten the paper so there's a point at the top.

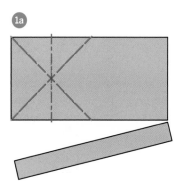

1a

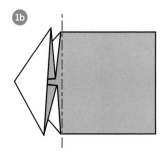

1b

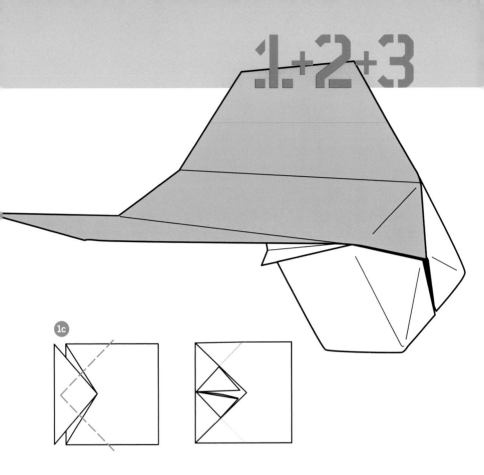

c. Turn over, and fold the triangle down along its bottom edge. Fold each entire top corner down along the centerline and unfold just the back corners. Leave the front corners folded down.

d. Invert the crease you just made on the back corners, so they lie underneath and inside, making the top pointy again.

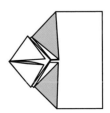

Paper Water Bomber

Continued

2. **Fold the cargo bay.**

    a. Fold the side of each flap in front to touch the centerline.

    b. Fold the bottom point of each flap up to touch the same point.

    c. The tricky part: Fold the bottom points up again diagonally, and tuck them inside the pockets you made in the previous step (see 2d).

    d. Fold each side corner of the diamond back to touch the centerline behind. Fold the top and bottom corners in front to touch the centerline. Unfold all four folds.

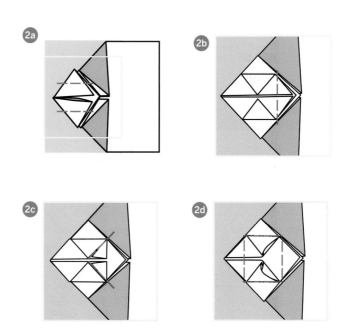

Ewan Spence (ewanspence.com) is a podcaster, blogger, reporter, new media junkie . . . generally a hoopy frood who's fun to be with.

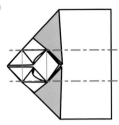

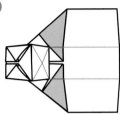

e. Crease the wings to line up with the edges of the cargo section.

f. Blow into the nose of your plane while gently pulling up to inflate the cargo section into a cube.

3. Bombs away!

a. Fill the cargo section with water, pouring carefully into the hole in front.

b. (Optional) Wait until no one is looking.

c. Chuck the bomb at your target or victim of choice.

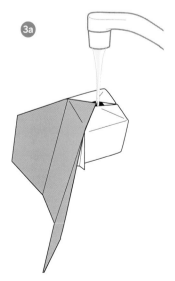

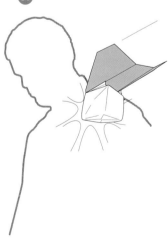

High-Pressure Foam Rocket

By Rick Schertle ■ Illustrations by Damien Scogin

YOU WILL NEED

Foam pipe insulation, ½" inside diameter You can build 8 rockets with a 6' piece (instructions here are for one rocket).

Foam sheet, 2mm thick, 9"×12" available at craft stores or online

Zip tie, 8"

Duct tape Fun colors are now available.

Scissors

Hot glue gun and glue

Packing tape, clear (optional)

CALLING A ROCKET THAT SPRINTS OVER 100 feet into the air a "toy" might be a bit of a stretch. Toy or not, this rocket really packs a punch. Fly it using the Compressed Air Rocket Launcher from the Maker Shed at makershed.com/products/compressed-air-rocket-launcher-v2-1 or a DIY stomp rocket launcher (see makezine.com/go/stomplauncher). Find all things air rockets at airrocketworks.com.

1. Cut the foam and cinch with a zip tie.

Cut a 9" section of foam pipe insulation. Wrap the zip tie ½" from one end and cinch it tight so no air will escape. Trim the excess foam above the cinched-off end.

2. Apply duct tape to foam rocket body.

Criss-cross 2 pieces of duct tape over the cinched end. Now cover the entire foam section with duct tape spanning the length. Three overlapping pieces of tape should do the trick.

3. Cut and attach foam fins.

Cut a 4"×1½" rectangle from the foam sheet, then cut it diagonally to make 2 fins. Repeat. Using generous amounts of hot glue, attach 3 of the fins, spaced evenly, onto the foam rocket body tube. If the glue doesn't stick well to the rough duct tape, wrap 2 strips of clear packing tape over the duct tape around the bottom of the rocket to create a smooth surface for gluing the fins.

Use It.

Pressurize the compressed air launcher for 45psi–65psi, and launch. (For a lower pressure/altitude launch, use a stomp launcher.) When the duct tape finally fails with a spectacular blowout, just apply more duct tape over the blown section and keep flying! ⚡

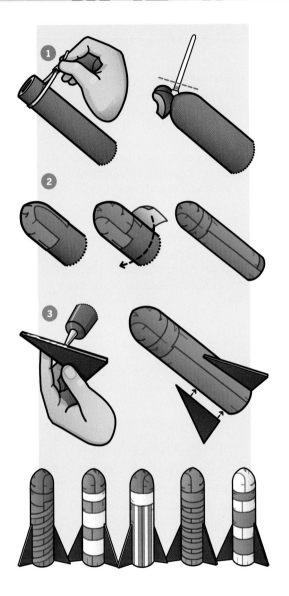

Rick Schertle teaches middle school in San Jose and spreads the Maker spirit through after-school classes. His wife and two young kids provide a constant flow of creative inspiration.

Remote-Controlled Camera Mount

By Ben Wendt ■ Illustrations by Julian Honoré/p4rse.com

YOU WILL NEED

R/C car with a relatively large base
Lightweight camcorder
Drill with ¼" bit
¼-20 threaded bolt with matching nuts, washers, and wing nut

WHAT DO YOU GET WHEN YOU COMBINE A CHUNKY REMOTE CONTROL TOY CAR with a lightweight camcorder? You get a street-level action cam that captures video on the move! I came up with this quick and easy mashup for kicks, and have had lots of fun with it. I hope you will, too.

1. Drill the chassis.

Remove the plastic body from the car and drill a ¼" vertical hole through the chassis.

2. Fasten the bolt.

Push the bolt up through the hole, securing it with nuts and washers above and below. Then put a wing nut at the top, facing upward.

3. Attach your camera.

The ¼-20 bolt fits most camcorders' standard tripod mounts. Screw your camera on, point it in your chosen direction, then secure it by tightening the wing nut.

Ben Wendt is a math and computer science dork from Toronto.

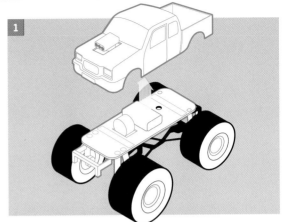

Use It.

It's really that simple! I've been using the camera to do Hale's Tours of my neighborhood. If you're unfamiliar with the history, Hale's Tours and Scenes of the World debuted in 1905.

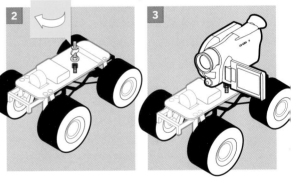

Charles Hale used to strap a camera to the front of a train going through a particularly interesting route, and film it. People would then pay to see simulated train rides through exotic locales in this fashion. As an homage to my community and to film history, I've been re-creating this experience using the remote-controlled camera mount. ✍

Origami Flying Disc
By Cy Tymony ■ Illustrations by Tim Lillis

YOU WILL NEED
One 8½"×11" piece of paper
Scissors
Transparent tape

UNDERSTAND BERNOULLI'S PRINCIPLE OF FLOWING FLUIDS AND GASES WITH A PAPER FLYER.

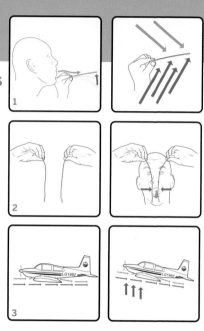

1. Demonstrate Bernoulli's principle.

Cut a paper strip ½"×4". Hold the paper strip just below your lips, and blow above the strip (Figure 1). The paper will rise!

This occurs because of Bernoulli's principle—fast-moving air has lower pressure than non-moving air. The still air below the strip has higher pressure than the moving air above, so it pushes the strip upward.

Try another quick test: cut 2 paper strips ½"×4". Hold them about 2" apart and blow air between them (Figure 2). You expect them to blow apart, but they actually come together. Bernoulli's principle is working here because the faster-moving air between the strips has lower pressure than the air outside them, which therefore pushes them together.

The top of an airplane wing curves upward and has a longer surface than the bottom. When the plane moves, the air moving along the top must travel farther and faster than the air moving past the straight bottom. The faster-moving top air has less pressure than the bottom air. This provides lift (Figure 3).

2. Make an origami flying disc.

Now use your new-found Bernoulli principle knowledge to make a flying disc using only paper and tape.

Cut eight 2"×2" square pieces of paper (Figure 4). Fold the top right corner of one square down to the lower left corner (Figure 5). Then fold the top left corner down to the lower left corner to create a small triangular pocket (Figure 6). Repeat these two folds with the remaining 7 squares (Figure 7).

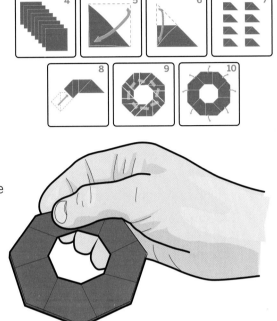

Insert one paper figure into the left pocket of another (Figure 8). Insert the figures into each other until they form an 8-sided disc. Hold the disc firmly together while you apply tape as required (Figure 9).

3. Fly it.

Turn the disc over and toss it like a Frisbee. You'll see that it glides a bit, but then drops rapidly. The reason is that the top and bottom surfaces are straight.

Now bend down the outer edges to form a curved lip (Figure 10). This should produce the Bernoulli effect.

Throw the flyer with a quick snap of your wrist, and it should stay aloft for a much greater distance. ◪

Cy Tymony (sneakyuses.com) is a Los Angeles-based writer and is the author of *Sneakier Uses for Everyday Things*.

Two-Cent Wobbler

By George W. Hart

YOU WILL NEED

2 identical coins or other disk-shaped objects that a slot
can be cut into, such as a CD

Dremel tool or saw to cut slots

NOTE: In some countries it is illegal to damage coins.

I RECENTLY RETURNED HOME FROM JAPAN with some leftover coins and decided to make them into "wobblers." Two orthogonally interlocked disks will roll together with an amusing left-right wiggle. If the spacing between the disks is just right, the wobbler's center of gravity remains at a fixed height so it will wobble down the slightest incline. It is surprisingly addictive to roll these and race them, and you can't beat them as a cheap gift!

1. Cut slots in any 2 identical coins.

With a cutting wheel on a Dremel or other rotary shaft tool, cut a small radial slot in each coin the same width as the coin's thickness. Hold each coin in a vise while you cut, and wear a mask so you won't breathe in any metal dust.

2. Press the 2 coins together.

Tap them gently with a hammer or apply gentle pressure with a vise. If the slots are not too wide, they'll hold together without any solder or glue.

3. Wiggle your wobbler.

Put the wobbler on a smooth surface and watch it wiggle away.

Going Further

After you make a few, you'll want to make mathematically ideal wobblers, in which the center of gravity remains at a constant height as it rolls. For this, the center-to-center distance must be the square root of 2 times the radius (d=√2r), so the slot length should be 29% of the radius.

I've since learned of several earlier discoveries of this shape, going back to the mid-1900s, when the designer Paul Schatz used what he called the "Oloid" as part of a paint stirring machine.

Physicist A. T. Stewart first observed that the center of gravity stays at a constant height if the center separation is √2 times the radius; see his paper "Two Circle Roller" in *American Journal of Physics*, Volume 34, 1966, pages 166–167. Search the web for "two circle roller" to read more interesting papers about these toys. ◪

George W. Hart is a sculptor and a professor at Stony Brook University. View examples of his work at georgehart.com.

1.2.3

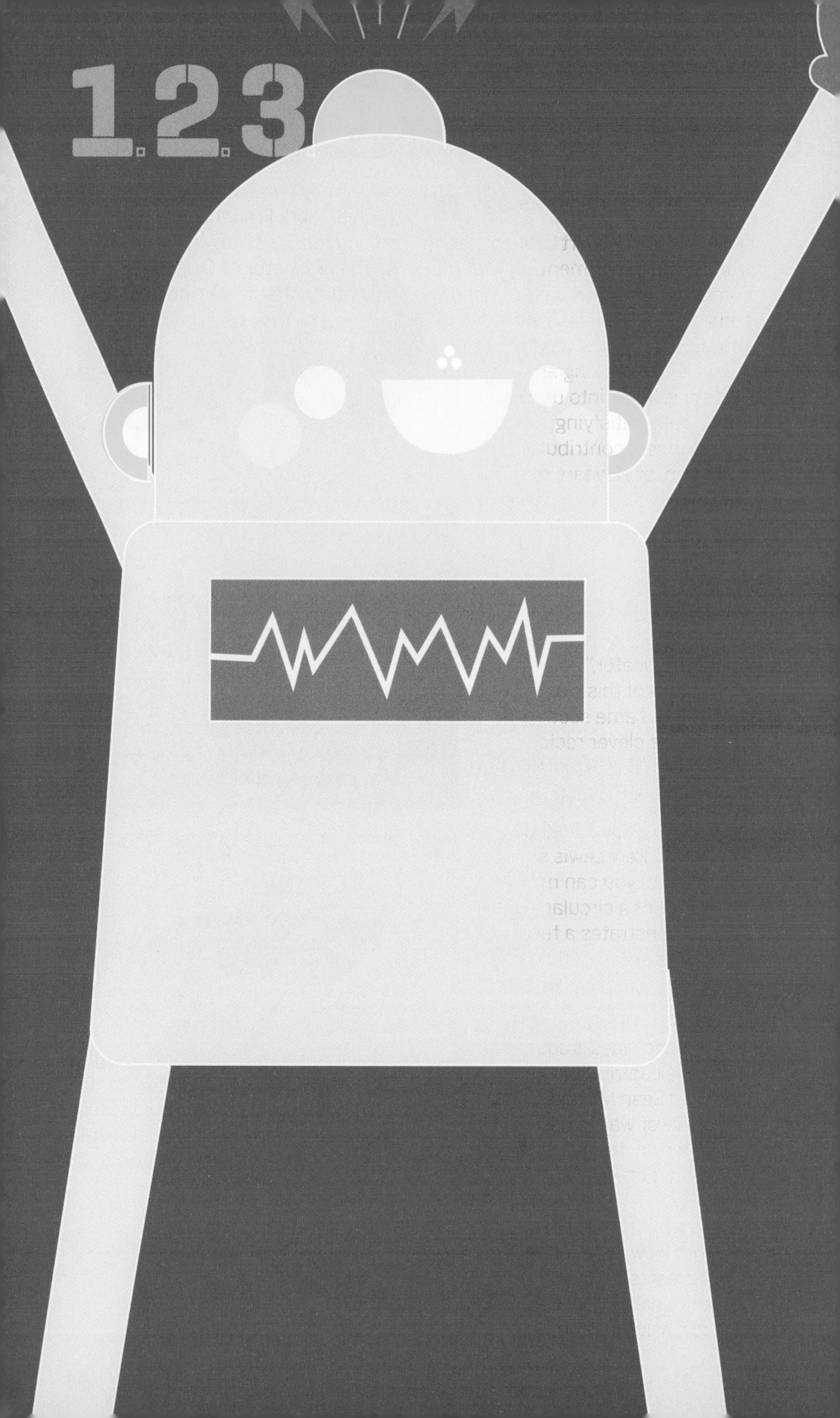

Part II: Arts and Crafts

RILEY MULLEN'S "MINI SPIN ART MACHINE" SPRANG FROM HIS LOVE of taking apart broken electronics and experimenting with the components. In this case, he had a computer fan and a 9V battery; add some pens and the next thing he knew he was making art. "Turning electronic 'junk' into unexpected fun is really satisfying," says Riley. The youngest contributor to this collection, at 12 years of age Riley is currently taking junior college classes and planning to be a chemist someday.

Cy Tymony introduces a drawing device from the 1600s called the pantograph with his "Mechanical Image Duplicator," an easy-to-make DIY version of this classic drawing tool. In this same section, Cy demonstrates a clever recipe for turning cow juice into moldable material in "Sneaky Milk Plastic."

With the "Cut-and-Fold Center Finder," Andrew Lewis shows you a clever tool you can make to find the center of a circular surface. He also demonstrates a technique for turning an ordinary tin can into an antique-looking container in "Tin Can Copper Tan."

Along similar lines, frequent *Make:* magazine contributor and technical editor Sean Michael Ragan outlines a clever way to use etchant to put your own designs on ordinary bottles in "Label-Etch a Glass Bottle." Meanwhile, Jason Poel Smith explains how to add fun to your kids' school lunch in "How to Tattoo a Banana." Jason creates a weekly how-to project series called "DIY Hacks and How Tos" for makezine.com.

Make: magazine founding editor-in-chief Mark Frauenfelder has an affinity for DIY stringed instruments. As the originator of *Boing Boing*, it's only fitting that Mark should contribute a project called "Boing Box" to the current collection.

UCSC robotics engineering student Paloma Fautley loves tinkering around in the kitchen, sewing room, tool shed, and at the electronics bench. In "Duct Tape Double" she shows you an easy and inexpensive way to make a mannequin to use for sewing and costuming projects.

Sindri Diego is a gymnast and an acrobat in Iceland's "first and only circus." Having learned about the artistic effect of blurring used in photography inspired him to create his own simplified approach and share it with you in "Bokeh Photography Effect."

Zitta Schnitt is a design professional whose office is located in Vienna, Austria. In "PET Bottle Purse," she describes an aesthetically attractive way to reuse beverage containers made from polyethylene terephthalate (PET).

Make: magazine editor Craig Couden shows you how to add some *Tron*-inspired fashion flair in "Simple Light-Up Hoodie." Morten Skogly's "Wind-Triggered Lantern" teaches you a neat trick for creating magical beauty that will turn your backyard into a midsummer night's dream.

Get ready to set your artistic impulses free! ◪

Mini Spin Art Machine
By Riley Mullen ■ Illustrations by Julie West

YOU WILL NEED

Computer fan 5V–12V
Battery 9V
Toggle switch two-terminal RadioShack
    #275-0022
Alligator clip leads (3)
Cardboard box
Double-stick foam tape
Masking tape
Scissors
Paper
Colored markers

THIS PROJECT CAME FROM MY LOVE OF TAKING APART BROKEN ELECTRONICS and experimenting with the components. I had an old computer fan, a 9V battery, pens, and the gift of boredom. Next thing I knew, I was creating interesting art! Turning electronic "junk" into unexpected fun is really satisfying.

1. Mount the fan and switch.
Attach the fan to the box top with double-stick foam tape. » To mount the switch, make a small hole in the box. » Remove the mounting nut from the switch, push the switch lever through the hole from the inside of the box, and replace the mounting nut to secure it in place.

2. Wire the fan, switch, and battery.
Make another small hole to push the fan wires through to the inside of the box. » Using one alligator clip lead, connect the fan's black wire to the negative (larger) terminal on the battery. » Using another alligator clip lead, connect the fan's red wire to one of the switch terminals. » Use the last alligator clip lead to connect the unused switch terminal to the positive (smaller) terminal on the battery.

3. Add paper and color.

Cut paper circles smaller than the diameter of your fan. » Attach a paper circle to the center of the fan, using a loop of masking tape on the back of the paper. » Turn the fan on and gently apply a colored marker to the spinning paper. Enjoy making colorful designs! ▪

Riley Mullen is a young Maker who is endlessly fascinated with electronic components, physics, and engineering. He enjoys reading, tinkering, creating things with household stuff, and hanging out with other Makers at his local makerspace.

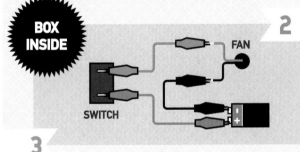

Cut-and-Fold Center Finder

By Andrew Lewis ■ Illustrations by Julian Honoré/p4rse.com

YOU WILL NEED

Thick cardboard cut to 10" square
Utility or craft knife
Pencil, ruler, tape

FINDING THE CENTER OF A CIRCLE IS EASY when you have the right tool. This cut-and-fold cardboard center finder is ideal for all those fiddly measuring jobs.

1. Measure and mark.

On your 10" square of cardboard, use the ruler and pencil to mark a diagonal line from the bottom left corner to the top right corner. Mark a vertical line ½" from the left side, a horizontal line ½" from the bottom, and another horizontal line 1" from the bottom.

2. Cut.

Using your utility or craft knife, cut out a ½" square from the bottom left corner of the card, using your ½" pencil lines as a guide.

Next, cut a triangle out of the card. Starting at the top right corner, cut along the diagonal line until you reach the 1" horizontal line at bottom left, then cut along the 1" line to the right edge of the card.

3. Fold and tape.

Use your knife to lightly score along the ½" border lines so that the card will fold easily, then bend the left and bottom ½" borders 90° to form a corner. Tape the corner to hold the cardboard edges at 90°.

Andrew Lewis is a keen artificer and computer scientist with interests in 3D scanning, computational theory, algorithmics, and electronics. He is a relentless tinkerer who loves science, technology, and all things steampunk.

Use It.

To mark the center of a circular object (for example, the top of a paint can), place your center-finding tool so that its folded edges touch the object's outside edges.

Now draw a line across the object, using the inside diagonal edge of your center finder as a guide. Rotate the object 90° and draw a second line. The point where the 2 lines intersect is the center of the circular object. ◪

NOTE: The maximum diameter of the circular object is roughly equal to twice the center finder's outside edge length—so your 10" center finder can be used with objects up to about 20" in diameter. To make a larger center finder, just follow the same instructions but use a larger square of cardboard.

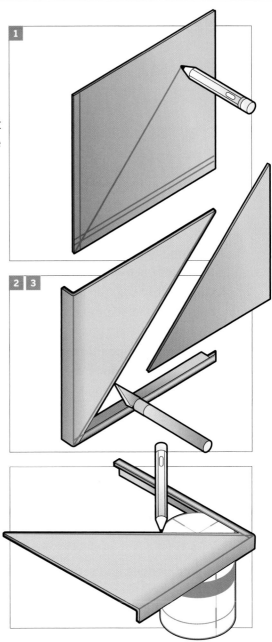

1

2 3

Mechanical Image Duplicator
By Cy Tymony ■ Illustrations by Alison Kendall

YOU WILL NEED
Thick white cardboard
Pencils **(2)**
Paper
Paper clips **(4)**
Paper clip boxes **(2)**
AA battery or other small weight
Marking pen
Transparent tape
Scissors

BEFORE CHESTER CARLSON INVENTED PHOTOCOPYING, inventors engineered various mechanical devices to replicate images. With a few everyday items found in the home, you can make a *pantograph*, an image duplicator that allows you to use one pencil to trace an image while another pencil follows its path in parallel to produce a near-identical copy.

1. Cut out and position cardboard strips.
You'll need 4 cardboard strips. Cut 2 strips measuring 2"×4" and another pair 2"×8", as shown in Figure 1. Place the 2 pairs of strips at right angles to each other, with the smaller pair lying on top of the larger pair.

2. Link cardboard strips with paper clips.
Cut 4 holes in the strips and slip 3 paper clips into them, as shown in Figure 2. Bend up the end of another paper clip, as shown, and tape it to the top of a paper clip box.

3. Add pencils and secure to table.
Cut 2 holes in the image duplicator strips large enough for 2 pencils to fit snugly and stand erect, as shown in Figure 3.

Turn the cardboard strips over and slip the hole at the end of the left-hand strip over the paper clip that's taped to the top of the paper clip box.

Place a second paper clip box under the image duplicator where the 2 large strips meet, to keep it level. To ensure that the drawing pencil (B) presses against the paper properly, you can add weight to the cardboard strip by taping a AA battery underneath it.

Use It.

Place the original image under pencil A, and a blank sheet of paper under pencil B. Trace the original design with pencil A. Pencil B will follow along, drawing the image on the paper.

Experiment with different lengths of strips to make larger and smaller copies of the original design. ◪

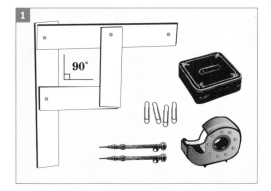

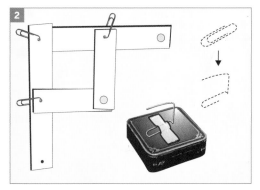

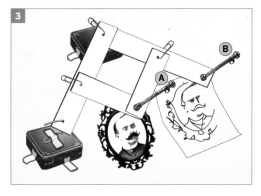

Cy Tymony (sneakyuses.com) is a Los Angeles-based writer and is the author of *Sneakier Uses for Everyday Things*.

Label-Etch a Glass Bottle
By Sean Michael Ragan ■ Illustrations by Julie West

YOU WILL NEED
Glass bottle with adhesive plastic label
Permanent marker
Hobby knife
Tweezers small
Scrap paper
Paper towels
Rubbing alcohol
Etching cream
Paintbrush
Safety goggles
Gloves
Sink

⚠ **caution:** Glass etchants are toxic and should be handled with care. Wear gloves and goggles and follow the label directions closely.

HERE'S A SIMPLE TRICK I DISCOVERED for etching designs on glass bottles using the bottle's label as a built-in resist.

1. Prepare the bottle.

This process requires a bottle with an adhesive plastic label. A sure sign that the label is suitable is that parts of it are

transparent. » Use a permanent marker to draw your design on the label. » Using your hobby knife, carefully cut around the edges of your design. Lift the edges of the cutout areas using the blade, and finish peeling off each positive cut using tweezers. » Wipe the cut stencil with a paper towel generously soaked with rubbing alcohol. This will remove residual ink and clean any remaining adhesive from the cutout areas. » To make sure the remaining stencil is firmly adhered to the bottle everywhere, wrap a scrap of paper around the bottle, over the label, and rub it briskly with the side of your marker.

2. Apply etching cream.

Generously daub etching cream over the exposed positive areas of your design using a brush. » Leave the etching cream in place 5 minutes, or whatever the instructions say, and then wash away all traces of the cream with plenty of warm water in the sink.

3. Remove the label and clean.

Using your hobby knife or just your fingernail, lift one corner of the label and peel it off. » Give the etched design one final cleaning with rubbing alcohol and a paper towel to remove any leftover adhesive. ▨

Sean Michael Ragan is technical editor of *Make:*. He's descended from 5,000 generations of tool-using hominids. Also he went to college and stuff.

Tin Can Copper Tan
By Andrew Lewis

COPPER-COAT A TIN CAN, TURNING IT INTO AN AESTHETICALLY PLEASING, REUSABLE CONTAINER.

1. Wearing protective clothing in a ventilated area, mix 9 parts of muriatic acid to 1 part hydrogen peroxide in a plastic container. Slowly add scraps of copper, which will react with the acid solution and turn it a blue-green color. When the reaction slows and the copper stops dissolving, remove the remaining pieces of copper. The blue-green component is cupric chloride, which reacts with tin and leaves a shiny new layer of copper in its place.

2. Thoroughly wash a tin can, removing all traces of glue and grease from the inside and outside. An abrasive plastic dish-cloth might be useful for this. Dry the can and gently place it into the acid solution. The can need not be completely submerged, but keep it turning to expose all parts to an equal amount of the solution. After a few seconds, the can will start to change color, and you should see a pale copper tan appearing within minutes.

3. Remove the can from the acid bath and rinse with water. Leave the can to dry naturally. Then you can finish it with clear acrylic spray, coat it with wax, or just let it oxidize to create a dark, neglected, antique look. ◪

Photography by Andrew Lewis

Andrew Lewis is a keen artificer and computer scientist with special interests in 3D scanning, algorithmics, and open-source software.

How to Tattoo a Banana
By Jason Poel Smith

DON'T JUST PLAY WITH YOUR FOOD—MAKE ART WITH IT.
I started tattooing bananas after seeing similar work by artist Phil Hansen. It is a great way to add some fun to packed lunches, and it doesn't ruin the food.

When you puncture or bruise a banana peel, the ruptured cells release chemicals that start to oxidize and turn brown. By using a fine-tipped needle, you can make detailed drawings on a banana peel. You can even use stencils to make a copy of your favorite picture.

1. Create a stencil.

Find a picture that you want to use and scale it to the appropriate banana size. Then print it out. Cut out the picture, leaving some blank paper around the edges. Next, use Scotch tape to attach the picture to your banana.

2. Outline the pattern.

Using a needle, poke holes through the paper along all the major lines. Try to keep the holes close together and as

Photographs by Jason Poel Smith

shallow as possible. To make this easier, try attaching a needle to a mechanical pencil. When you're done, remove the stencil. You should see a dotted outline of your picture.

3. Fill in the details.

Now you need to connect the dots and apply shading. Go back over all the major lines and make more dots to fill in the gaps. To apply shading, gently tap the surface to make very light scratches. The details will darken over time. It's art that you can eat! ▨

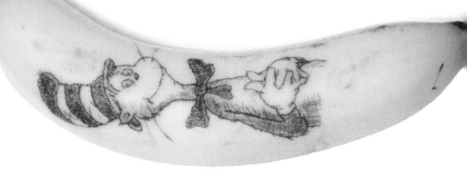

Jason Poel Smith is a lifelong student of all forms of making, from electronics to crafts and everything in between. He creates the "DIY Hacks and How Tos" video series for *Make:* at youtube.com/make.

Sneaky Milk Plastic
By Cy Tymony

YOU WILL NEED

Metal saucepan
Spoon **for stirring**
Measuring spoons
Strainer
Wax paper
Paper towels
4tsp white vinegar
1c milk

Optional:
Acrylic paint
Cornstarch
Small neodymium magnet

DON'T HAVE A 3D PRINTER TO MAKE PLASTIC PARTS?

Use moo juice instead. Ordinary cow's milk contains a protein called casein. When separated from milk by using an acid such as vinegar, casein becomes a moldable plastic material that can be used to create everything from glue to fabric to billiard balls. Make your own custom parts for projects!

1. Cook the milk with vinegar.

Pour 1 cup of milk into the saucepan and warm it to a simmer, not a boil, on the stove.

Next, add 4 teaspoons of vinegar to the milk and stir. After a few minutes you should see white clumps form. When you do, keep stirring a few minutes longer, then turn off the heat to allow the pan to cool.

2. Strain the casein from the milk.

Pour the milk through a strainer into a bowl to separate all the white clumps (this is the casein plastic material), and place them on a sheet of wax paper.

3. Dry and mold the casein plastic.

Dry the casein plastic material by blotting it gently with paper towels until it's dry.

Mold the plastic material into the shape(s) you prefer, and let it dry for at least 2 days. Once it hardens, you can color it with acrylic paint if desired.

> **TIP: If the casein is too runny to shape in your hands, next time add a pinch of cornstarch to the milk and vinegar mixture. This will make it hold together better.**

Going Further

Shape your sneaky milk plastic into a sneaky finger ring that will attract paper U.S. currency! Just hide a small, super-strong (neodymium) magnet inside your ring before it dries and hardens. When the ring is placed close to a legitimate folded bill, the bill will move toward the ring because of the iron particles in the currency's ink.

Cy Tymony (sneakyuses.com) is a Los Angeles–based writer and is the author of *Sneakier Uses for Everyday Things*.

Boing Box
By Mark Frauenfelder

YOU WILL NEED

A cigar box
6' length of ½"×¾" wood
8' of 20-gauge wire
Eye screws (2)
Wood screws
L-bracket
Scraps of wood
Drill, saw, and screwdriver

BUILD A FUN, ONE-STRINGED INSTRUMENT THAT PACKS A MIGHTY TWANG.

A 1951 book called *Radio and Television Sound Effects*, by Robert B. Turnbull, shows how to make a "boing box." (It's reprinted at bizarrelabs.com/boing2.htm.) I made a modified boing box using a wooden cigar box and some scraps I had around the house.

1. Drill resonator holes in the cigar box as shown. Screw an L-bracket to the neck, then screw the neck to the cigar box. To prevent structural failure, put a ½-inch-thick scrap of wood under the lid and drive the screws into it. This will keep the screws from pulling out when the wire is tightened.

2. Insert eye screws into the cigar box and the end of the neck. Use another wood scrap on the inside of the box for the eye screw. Tie and tighten wire to both eye screws. The wire should be tight enough to cause the neck to bow slightly. (I used an eyebolt with hex and wing nuts to make it easy to adjust.)

3. Pluck the string and gently shake the boing box to vary the pitch. Boing! ▨

Photography by Mark Frauenfelder

Mark Frauenfelder is the co-founder of Boing Boing, former editor-in-chief of *Make:* magazine, and author of several books.

PET Bottle Purse

By Zitta Schnitt ■ Illustrations by Tim Lillis

YOU WILL NEED

16oz PET bottles (2) such as soda bottles
9" zipper
Tape
Nylon thread
Vise optional
Hacksaw
Small pair of scissors
Thin needle
Thick, sharp needle
Pliers or seam ripper

FOLLOW THESE INSTRUCTIONS TO BUILD A COOL PURSE out of the bottoms of 2 PET bottles and a zipper. You can use the container as a coin purse or a storage box.

1. Cut the bottles.

Cut off the bottoms of both bottles using the hacksaw, while holding the bottle in place with a vise. Then trim the burrs and any excess material with the small scissors.

2. Punch the holes.

Download, print, and cut out the hole-punching template found at cdn.makezine.com/make/25/M25_123_Bottlepurse_Hole_ Template.pdf. Tape the template, dots facing up, to the outside of the bottle, the top edge of the template flush with the cut bottle edge.

   Holding the thick needle with a pair of combination pliers or vise-grip pliers, punch holes in the bottle, using the dots on the template as a guide. (You can also use a seam ripper to punch the holes.)

   When you're done, move the template to the other bottle bottom and repeat the hole-punching process.

3. Attach the zipper.

Unzip the zipper before sewing it into both bottle bottoms. With the thin needle and thread, begin stitching 1¼" below the zipper's top end, leaving it loose for now (later you'll overlap it with the zipper's bottom, and stitch them both down).

Stitch the zipper around one bottle bottom, and then around the other (inverted) bottle bottom.

To finish, overlap the zipper's bottom end with its top, and sew both ends to the bottle bottoms through the remaining holes. ◨

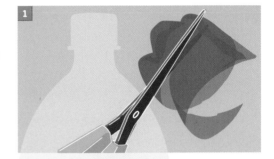

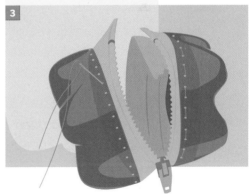

Zitta Schnitt, born in Vienna, spent her childhood in Austria and Hungary. In art school she focused on textile design. Now she studies industrial design at the University of Applied Arts in Vienna, with a special focus on tableware and lighting design. zittaschnitt.com

Duct Tape Double
By Paloma Fautley ■ Illustrations by Julie West

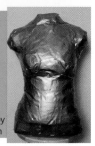

YOU WILL NEED

Duct tape
Old shirt you don't mind ruining
Newspaper or old clothes for stuffing
Eye screws
Scissors
Plastic wrap optional

Photography by
Gunther Kirsch

THE DUCT TAPE DOUBLE PROJECT is a quick and easy way
to make a custom mannequin. It uses cheap materials you
can find around the house to create a dimensionally accurate
replica. Use your new double to craft custom garments that
fit perfectly every time, or step it up a notch and venture into
well-tailored wearable technology.

1. Prepare the model.

Use a tight-fitting shirt that covers the area that you want to
replicate. If you want to add length beyond the shirt, cover the
additional area with plastic wrap. Make sure that the model is
comfortable and can breathe easily.

2. Wrap it up!

Wrap the model with duct tape. Use the patterns provided as
a general guideline. Don't pull too tightly or the double won't
accurately represent your true size.

   Cover every desired area with 2–3 layers of duct tape so it is
nice and sturdy.

3. Remove and stuff.

Cut the duct tape double off your model carefully, slicing up
the back and making sure to not cut the model in the process.
Carefully slide the double off. Be gentle and try not to lose any

features while doing so. Stuff the double with old clothes, newspaper, or other materials to make it sturdy. Do not overstuff or you will lose the shape. Tape up the openings, and you're finished! ◪

1

FEMALE MALE

2

CUT IT & STUFF IT

3

Paloma Fautley is an engineering intern at *Make:*. She is currently pursuing a degree in robotics engineering and has a range of interests, from baking to pyrotechnics.

Simple Light-Up Hoodie

By Craig Couden ■ Illustrations by Julie West

YOU WILL NEED

Sweatshirt with large hood
EL tape strip, about 40" Adafruit #415
Pocket-sized EL inverter 4×AAA or 2×AA
EL wire extension cord Adafruit #616
Velcro enough to match the EL tape
Batteries rechargeable
Sewing machine or needle and thread
Glue optional

TRON: LEGACY **WAS A DECENT MOVIE,** but what I was really wowed by were the luminescent costumes. I've been playing around with light-up clothing ideas ever since, and the simplest idea uses EL tape and some Velcro to approximate Jeff Bridges's glowing druid look from the film. It's not remotely screen accurate (for you purists out there), but it is a great way to wow your friends during a night out.

1. Sew Velcro to hood.

Cut a length of Velcro equal to the length of the EL tape and hand or machine stitch the soft side onto the inside of the hood near the edge. You may want to remove the drawstring first.

2. Attach Velcro to EL tape.

Attach the prickly "hook" side of the Velcro to the back of the EL tape. (Mine came with an adhesive backing, but you may have to glue it.)

3. Juice it.

Attach the EL tape to the hood. Connect the EL tape to the extension cable, then the inverter. Turn it on, stick the inverter in your pocket, and show that you, too, fight for the User!

Going Further

To achieve a longer wear time, use 2 shorter lengths of EL tape that meet at the apex of the hood, each with its own 2×AA inverter.

When not obsessing over his new couture EL fashion line, Craig Couden is an editorial assistant for *Make:* magazine.

Wind-Triggered Lantern
By Morten Skogly ■ Illustrations by Alison Kendall

YOU WILL NEED

LED
3V button battery
Pencil, ruler, tape
Something springy made of metal
Feather
Thread
Soldering iron and solder, or tape
Carpet knife
Strong wire
Battery holder (optional) recommended
Glass jar with lid (optional) for weather protection

CREATE A LITTLE MAGIC IN YOUR YARD with this flickering garden lantern triggered by the wind, made with spare parts you probably have lying around your house.

1. Attach the LED to the battery.
Solder one of the LED's leads to the battery holder (Figure 1). I got the battery holder for the flat 3V button cell battery from an old PC that I've been scavenging parts from. (It's the battery that powers the internal clock, and I guess every PC has one.) You don't need a battery holder at all—you could just tape one of the LED's feet to the battery—but a battery holder makes things easier.

> **NOTE:** Remember to test the LED first, so you know you're attaching the correct lead to the correct side of the battery.

2. Make the flickering mechanism.
Solder a flexible piece of metal to the other side of the battery holder. I happened to have a long, thin spring from the CD-ROM drive of an IBM ThinkPad I took apart a while back; it works great. Another option could be a copper thread or wire, as thin as possible, or a piece of guitar string. Then bend the

unsoldered lead of the LED so it curves around the spring without touching it (Figure 2).

Attach a feather to the spring with a piece of thread. When the feather moves in the wind, it pulls on the spring, which touches the foot of the LED and closes the circuit—which equals blinking!

3. Hang it in the garden.

For weatherproofing, cut a slit in the lid of a jar and put the feather through it (Figure 3). Fiddle with it until the mechanism moves freely. Use 1yd or so of strong wire to wrap around the jar, to make a "harness" and a handle. Then go out and hang it in the garden (or run around with it, giggling, like I did). Possible improvements: Add a solar cell and battery. Create a prettier casing, perhaps using beeswax? Or maybe even add sound! ◪

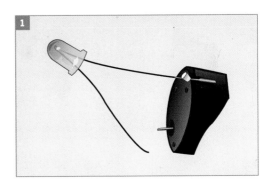

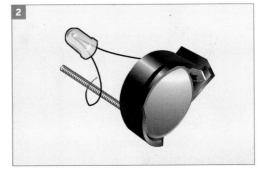

Photograph by Morten Skogly

Morten Skogly is a cheerful man, even in the dark. He makes things to impress his wife and kids. See more of his DIY stuff at pappmaskin.no.

Bokeh Photography Effect
By Sindri Diego ■ Illustrations by Damien Scogin

YOU WILL NEED

Paper thick and black
Camera lens 50mm
Camera
Compass
Ruler
Scissors
X-Acto knife

For more on bokeh effects, visit our website at:
makezine.com/projects/bokeh-filter/

Or visit our YouTube channel at:
www.youtube.com/watch?v=YZjV94FbjRE

BOKEH COMES FROM THE JAPANESE word for "blur." In photography, the bokeh effect has to do with the aesthetics of out-of-focus areas of the picture.

With this project, you can create an effect that makes the out-of-focus lights in your pictures appear any shape you want.

1. Measure and cut a paper disk.

Set your compass to measure 25mm between the spike and the pencil, and draw a circle 50mm in diameter.

Cut out the circle, leaving a little tab of paper somewhere on the edge to use as a handle. If you use the paper wisely you can make many bokeh disks from one sheet.

2. Cut your shape out of the center.

Draw your shape, centered, on the 50mm paper disk. It's best if the shape is not too complicated.

Using the X-Acto knife, cut out the shape you drew. Be careful to not cut yourself.

Sindri Diego is a 19-year-old Icelandic multimedia design student, gymnastics champion, and coach. He is taking his first steps in photography and loves to try new things with it.

3. Start making cool bokeh effects.

Position your bokeh disk in front of the lens. You can probably fix it in place using the threading that's intended for attaching filters. It doesn't matter how close the paper is to the glass lens, but it has to cover the whole lens.

Adjust your camera to the lowest aperture setting and start shooting. Remember, you want to have the lights out of focus to get the effect. Be as creative as you can—if you have a digital camera, you can always delete the images you don't like.

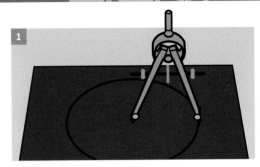

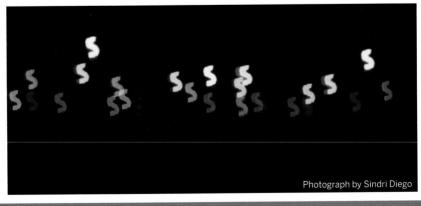

Photograph by Sindri Diego

1+2+3

Part III: Science and Electronics

CY TYMONY HAS CONTRIBUTED MORE ARTICLES TO *MAKE:* MAGAZINE'S 1+2+3 FEATURE than any other writer. More than a dozen published books, with titles like *Sneaky Uses for Everyday Things*, attest to Cy's restless imagination and boundless creativity. His 10 projects in this section range from switches and batteries made of everyday things to a clever hack for turning an ordinary AM/FM radio into an aircraft band receiver.

Mark Frauenfelder makes an ingenious little "bot" with a motor and a mint tin in his "Vibrobot" project. For more robot fun, check out Steve Hoefer's "Dizzy Robot." A former Iowa farm boy who was making (and breaking) things as soon as he figured out which end of the tool to hold, Steve is today the writer, inventor, and creative problem solver behind Grathio Labs in Las Vegas.

Casey Shea's journey to becoming a Maker began in 2011 when *Make:* founder Dale Dougherty approached him to start an experimental class called Project Make at the high school where he taught math. Today he's not only teaching students but also helping other teachers master the new tools and approaches of Maker education. Casey's "Springboard" project resurrects a 50-year-old tool for teaching electronics and then adds a modern update that's new to this collection.

Samuel Johnson and AnnMarie Thomas demonstrate that electronics can take any shape you can imagine in "Sculpting Circuits," a fun project for kids to experiment with LEDs. For even more fun, check out "LED Throwies" by Graffiti Research Lab—a great way to decorate your environment!

Doug Watson provides practical advice for mounting your GPS device in "Rear-View Power Socket," while Sean Ragan presents a useful hack to keep your Arduino board safe in "Board Feet." Freelance magazine writer and acclaimed suspense novelist Phil Bowie has a nifty trick for playing a record album with a paper clip and a packing peanut, which you can read about in his article "Paper Clip Record Player."

Photographer Danny Osterweil and Photojojo.com present an entertaining project for turning your smartphone into a DIY slide projector. In "Smartphone Signal Generator," software developer Jacob Beningo explains how to use your phone to generate signals to test speakers and other devices, generate audio tones, or just to experiment with circuits. Jacob enjoys designing practical tips and tricks for engineers and Makers.

Did you know that a piece of iron loses its attraction to a magnet when heated? Science experimenter and writer John Iovine's "Curie Engine" shows you how to use that principle to create a simple engine. "Sound Sucker," by bestselling author William Gurstelle, employs common kitchen materials to create a device that seems to cancel out certain audible frequencies. Try it yourself! ◪

Vibrobot
By Mark Frauenfelder

WHEN MY 3-YEAR-OLD DAUGHTER dropped the $1 battery-powered fan I bought her, the plastic case cracked, ruining it. I promised her I'd make something even better using the fan's motor. I'm a fan of Chico Bicalho's wonderful windup toys, so I made a robot inspired by his designs. I call mine the Vibrobot, and you can make one in a couple of hours or less.

1. Prepare the candy tin.
Sand the paint off the tin, if you wish. Punch 2 holes through the bottom of the tin on either end, using a hammer and a Phillips screwdriver. You'll use these holes to attach the legs. Punch a hole through the lid near one end. This hole is for routing the wires.

2. Make the legs.
Snip off 2 long pieces of wire from a coat hanger and bend each into a V-shape. Bend the tip of the V into a right angle, and then bend a little "foot" at each end (Figure A). Attach the legs to the holes in the tin using bolts, nuts, and metal washers (Figure B). Add a dollop of hot glue to each foot to give them rubber tips.

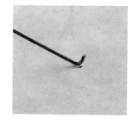

3. Install the motor.

Push a paper clip through one of the plastic flat washers, and attach the washer to the spindle of the motor. Solder 2 wires to the 1.5V battery, insert the battery in the candy tin, and thread both wires through the hole in the lid. Solder one wire to a lead on the motor, and solder a third loose wire to the other motor lead. Put 2 plastic flat washers between the motor and the candy tin, and secure the motor to the tin using a cable tie.

To operate the Vibrobot, twist the loose battery wire and the loose motor wire together (you can also

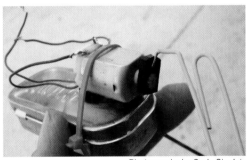

Photography by Carla Sinclair

solder an alligator clip to one of the wires for a switch). Experiment with the critter by gently bending the paper clip and legs into different shapes and observing the effects. Watch a video at makezine.com/projects/make-10/vibrobots/. ◪

Mark Frauenfelder is editor-in-chief of *Make:*.

Magnetic Switches from Everyday Things

By Cy Tymony ▪ Illustrations by Mark Frauenfelder

YOU WILL NEED

Magnet

Paper clip

Aluminum foil

Tape or foam

Cardboard

Wire

LED or buzzer

3V watch battery or equivalent

Ring optional

Battery-operated toy optional

X-10 universal
  interface optional

Appliance module optional

CONTROL MANY DEVICES FROM afar with the magnetically sensitive Sneaky Switch.

1. Make a magnetic activator.

You'll want a strong magnet to activate devices from at least an inch away. Tiny rare earth magnets can be found in most micro radio-controlled cars, and in the packaging of some hearing aid batteries. Glue a magnet to the face of a ring or a wand, or affix it to some object so that when it's near the switch, or moved away, it will cause the desired effect.

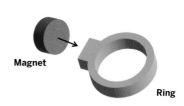

Magnet

Ring

2. Make a Sneaky Switch.

In this magnetic switch, the paper clip lies across a "spring" of rolled tape, one end hovering just above the aluminum foil and the other end taped down.

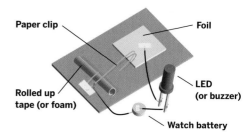

Paper clip

Foil

Rolled up tape (or foam)

LED (or buzzer)

Watch battery

(A small piece of foam can also be used as the spring.) When a magnet passes over the switch, it tugs the clip to touch the foil, completing the circuit. Connect the switch to a 3V watch battery to light an LED, buzzer, or other low-current devices and toys.

3. Connect switch to a relay.

Your magnetic switch can be attached to a relay to control devices that need higher current. Mount your switch and relay behind the dashboard to secretly activate a cut-off switch, alarm, or other car accessories. Or hook your switch to an X-10 controller and universal interface module to control a variety of appliances. Pretty sneaky!

Bonus: Detect counterfeit money.

Legitimate currency has iron particles in the ink. Fold a bill so half of it stands up vertically—if the top edge moves toward your magnet, it's the real deal. If not, phone the Secret Service! ◪

Cy Tymony (sneakyuses.com) is a Los Angeles–based writer and is the author of *Sneakier Uses for Everyday Things*.

Board Feet

Written by Sean Michael Ragan ■ Illustrations by Julie West

YOU WILL NEED

Development board with 3+ mounting holes such as Arduino Uno, BeagleBone Black, etc.

14mm × 8 mm recessed rubber bumper (4) Uxcell #A11120700UX0247

M3 × 10mm F/F threaded hex standoff (4) McMaster-Carr 92080A110

M3 lock washer (8) McMaster-Carr 91111A118

M3 × 8mm JIS pan head Phillips machine screw (8) McMaster-Carr 94102A103

M3 flat washer (8) McMaster-Carr 93475A210

Phillips head screwdriver

Electrical tape

Drill

File

WHEN WORKING WITH ARDUINO and other boards, most people just set the bare PCB down on the bench, hook up cables and components, and go. As long as your benchtop isn't made of metal, and you're careful, this can work fine. But it's not ideal. Clipped leads and other metal junk can short across the exposed solder points, causing erratic behavior and even damage. Give your board a lift by adding legs built from off-the-shelf hardware.

1. Make a special screw.

(Arduino only) The Uno REV3, Mega, and other Arduino boards have a minor design flaw in the mounting hole nearest the USB port—the nearby pin headers are too close to clear any standard screw head. To fix, just turn one of your screws tightly into one end of a loose standoff, wrap the standoff in electrical tape to prevent marring, and chuck it into a handheld drill or drill press. Fire up the drill, then use a small file (with one face taped to make it "safe") to turn down the screw head to 0.175"/4.4mm or so.

2. Mount the standoffs.

Pass four screws through the mounting holes from the component side. Slip a lock washer over the threads on the solder side, then add the standoffs and tighten down.

3. Install the feet on the standoffs.

Put a lock washer, then a flat washer over the threads of each of four screws before passing it through the recessed bumper from the bottom. Add a second flat washer on top of the bumper, then thread the screws into the free ends of the mounted standoffs. Tighten securely, and you're done! ↗

Check out step-by-step photos at:
makezine.com/projects/prop-up-dev-board-rubber-feet/

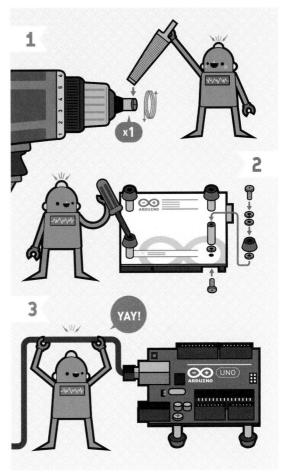

Sean Michael Ragan

Sean Michael Ragan (smragan.com) is a writer, chemist, and longtime *Make:* contributor. His work has also appeared in *ReadyMade*, *c't—Magazin für Computertechnik*, and *The Wall Street Journal*.

Batteries from Everyday Things

By Cy Tymony ■ Illustrations by Julian Honoré

YOU WILL NEED

Lemon or other acidic fruit
Nail, paper clip, or twist-tie
Heavy copper wire
Water
Salt
Paper towel
Pennies and nickels
Plate

NO ONE CAN DISPUTE the usefulness of electricity. But what do you do if you're in a remote area without AC power or batteries? Make sneaky batteries, of course! And once you know how to make sneaky batteries, you'll never again be totally out of power sources.

1. The fruit battery.

Insert a nail or paper clip into a lemon. Then stick a piece of heavy copper wire into the lemon. Make sure the wire is close to, but does not touch, the nail (Figure 1). The nail has become the battery's negative electrode and the copper wire is the positive electrode. The lemon juice, which is acidic, acts as the electrolyte. You can use other electrode pairs besides a paper clip and copper wire, as long as they're made of different metals.

The lemon battery will supply about ¼ to ⅓ of a volt of electricity. To use sneaky batteries to power a small electrical device, like an LED light, you must connect a few of them in a series, as in Figure 2.

2. The coin battery.

With the fruit battery, you stuck the metal into the fruit's electrolyte solution. You can also make a battery by placing a chemical solution between 2 dissimilar metal coins.

Dissolve 2 tablespoons of salt in a glass of water. This is your electrolyte solution.

Now moisten a piece of paper towel in the salt water. Put a nickel on a plate and put a small piece of the wet absorbent paper on the nickel. Place a penny on top of the paper. Next, place another moistened piece of paper towel on top of the penny, and then another nickel, and continue the series until you have a stack.

The more pairs of coins you add, the higher the voltage output will be. One coin pair should produce about $\frac{1}{3}$ of a volt. With 6 pairs stacked up, you should be able to power a small flashlight bulb, LED, or other device (Figure 3) when the regular batteries have failed. Power will last up to 2 hours.

Adapted with permission from *Sneaky Green Uses for Everyday Things* by Cy Tymony, Andrews McMeel Publishing, 2009.

Cy Tymony (sneakyuses.com) is a Los Angeles–based writer and is the author of *Sneakier Uses for Everyday Things*.

NOTE: If the LED does not light, reverse the connections on its leads.

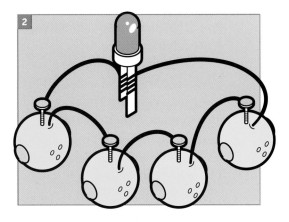

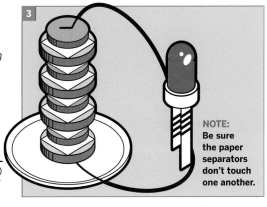

NOTE: Be sure the paper separators don't touch one another.

Springboard

By Casey Shea ■ Illustrations by Julie West

YOU WILL NEED

Plywood or soft wood, 4"×6"×¾"
Springs, 2" with loops on
  ends (2)
Springs, ½" with loops on ends
  (4–15)
Wood screws, ¾" (2 per spring)
Power drill
Battery (I used a 9V)
Snap battery connector 9V

I GET A VICARIOUS thrill watching young Makers' eyes light up like the LEDs in the conductive dough they're working with. But that feeling is too often followed by shared frustration at the increased difficulty when they switch to a solderless breadboard.

To ease the transition, I call back to duty the springboard, a 50-year-old relic that's difficult to find these days, but is easy and cheap to make. The layout mimics a breadboard, so by the time students reach the limitations of the springboard, they're ready to move onto a breadboard.

1. Design your board.

Lay out the springs (while not under tension) and mark where you'll attach them to the board. The number of springs depends on the complexity of the circuits you'll build. The most basic design has two vertical 2" springs on the left and right sides, with two rows of ½" springs running horizontally between them, two or three springs in each row. More complicated circuits require more rows. Space the springs so the legs of an LED can easily span two of them.

2. Screw down the springs.

Screw them down with your power drill, making sure that none of the screws are touching each other (finish nails are best for boards with closer spacing).

3. Energize!

Spread one of the 2" springs with a pointed object (like a multimeter probe) and insert the positive wire from the snap connector. Connect the negative wire to the other 2" spring. Complete the circuit by connecting individual components from spring to spring, using jumper wires when needed. Connect the battery to the snap connector. The springboard expresses the connected circuit visually and makes it much easier to practice using a multimeter at various locations. ◪

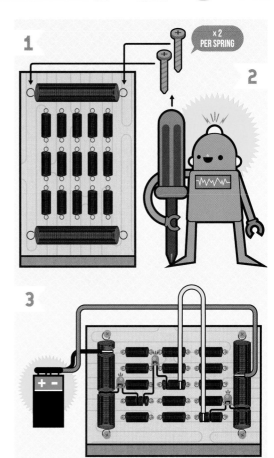

Gunther Kirsch

Springboard Update

MY ORIGINAL HORIZONTAL-SPRINGBOARD WAS based on a classic design. I've since created an alternative design using vertical springs.

The motivation for designing the vertical model was to make it easier to use alligator-clip jumper wires as well as regular solid or stranded wire without the need of tools. The horizontal model requires some sort of sharp object to spread the springs wide enough to insert the wire. The vertical springs can just be pushed to the side with a finger to make a big enough opening to insert wires. As with the horizontal model, this is used only to make simple connections and to give users a better understanding of the function of a breadboard. The photos don't show it, but we often put long springs horizontally along the sides to simulate the rails.

The board is made by drilling holes slightly smaller than the spring diameter. It's really a trial and error process based upon the size of the springs used. The fit needs to be pretty tight for the springs not to pop out of the holes. We use a laser cutter in 1/8'' plywood, but drill bits in any thickness of wood should work fine as well. A bit of epoxy or hot glue in the bottom of the hole (or on the back side of the plywood) helps the springs to stay in place.

Casey Shea teaches math and Project Make at Analy High School in Sebastopol, CA—the hometown of *Make:*. In addition to teaching students the skills of making, he is interested in sharing with educators the ways modern tools can be used to create custom instructional materials for their classrooms.

Easy Motor

By Cy Tymony ■ Illustrations by Dustin Hostetler

YOU WILL NEED

Two strong, metallic disc magnets approximately ¾" in diameter
   (larger in diameter than a AA battery)
6" length of stiff copper wire
AA battery
Needlenose pliers

MAKE A SPINNING MOTOR with a minimum of parts.

1. Make it.

First, if your copper wire is insulated, strip the insulation off. Bend the top of the wire into a hook shape. Press the tip into a sharp point with the pliers. Wrap the wire in a spiral form around the body of the battery, as shown in Figure 1. The wire should be just loose enough so it will not contact the battery case.

Next, place 2 metallic disc magnets on the battery's negative terminal. Adjust the shape of the wire so the top rests on the center positive terminal and the bottom just touches the side of the magnet. The length, when twisted, from the tip of the top hook to the bottom of the wire should be 2".

2. Try it.

Once the wire makes contact with the top battery terminal, it will spin. If it does not, carefully bend and adjust the wire so that it is free to move and make contact properly. To make the wire spin in the other direction, turn the magnets over.

3. Understand it.

When the wire touches the battery's positive top terminal and the side of the magnet, electricity flows through it. The wire

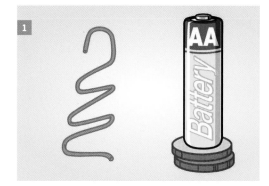

becomes magnetic and is repelled by the 2 disc magnets. When the wire moves away from the magnet, it disconnects from the battery and loses its magnetic field. It falls back to its original position and contacts the side of the magnet. Then, it connects to the battery through the metallic magnet and becomes magnetically charged again, and the cycle repeats.

Since the wire coil is suspended by its sharp tip on top of the battery's positive terminal, when it's repelled from and returns to the magnet, it rotates slightly. The cycle occurs so rapidly that it produces a circular motor motion. ◪

Cy Tymony (sneakyuses.com) is a Los Angeles–based writer and is the author of *Sneakier Uses for Everyday Things*.

Sculpting Circuits

By Samuel Johnson and AnnMarie Thomas

■ Illustrations by Julian Honoré/p4rse.com

YOU WILL NEED

For conductive dough:
  1c water
  1½c flour
  ¼c salt
  3Tbsp cream of tartar
  1Tbsp vegetable oil
  Food coloring (optional)

For insulating dough:
  1½c flour
  ½c sugar
  3Tbsp vegetable oil
  1tsp granulated alum
  ½c distilled or deionized water
  (check lab supply stores)

Assorted LEDs
4 AA batteries in a battery holder
Low-current DC motors

MAKING PLAY-DOUGH CREATURES IS fun, but making them with light-up eyes and moving parts is even more enjoyable. We thought it would be better still if we could make the circuitry out of the dough itself!

Most play dough is already conductive, but we needed a way to insulate the conductive dough. We came up with a sugar-based dough that works well as an insulator. It's pliable and resistant to blending with the conductive dough.

Rainy day and fidgety kids? Whip up both types of dough, gather some LEDs and batteries, and create your own menagerie of squishy circuit creations. Add a motor or two for sculptures with moving parts. Feeling adventurous? Play with the salt content of the recipes to vary their conductivity.

1. Make the conductive dough.

Reserve ½c flour, and mix the remaining ingredients in a medium-sized pot. Cook over medium heat, stirring continuously. The mixture will begin to boil and get chunky. Keep stirring until a ball forms in the center of the pot. Once a ball forms, turn off the heat and remove the dough to a lightly floured surface.

Slowly knead the remaining flour into the ball until you've reached the desired consistency.

Store dough in an airtight container or plastic bag. In the bag, water from the dough

 caution: The dough will be very hot! Flatten it and let it cool for a couple of minutes before handling.

will create condensation. This is normal. Just knead the dough after removing it from the bag, and it will be as good as new. Stored properly, it should keep for several weeks. If it dries out, just add a little more deionized water and knead it with some flour.

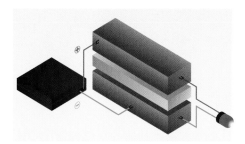

2. Make the insulating dough.

Mix the dry ingredients and oil in a pot or large bowl. Mix in 1 Tbsp of deionized water and knead; repeat until the mixture becomes moist and dough-like.

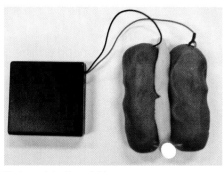

Photograph by Samuel Johnson

Remove the mixture from the pot or bowl, and slowly knead flour into it until it attains a firm consistency. You should use almost the entire ½c of flour.

NOTE: You probably won't need more than ¼c of deionized water, but have ½c ready just in case.

3. Make squishy circuits.

Insert the 2 leads from the battery pack into 2 pieces of conductive dough, separated by a lump of insulating dough (we recommend using food coloring to differentiate the doughs).

Insert an LED so its anode (long lead) is in the positive battery lump, and its cathode (short lead) is in the negative battery lump. It will light up! ◪

Samuel Johnson is from Blaine, MN, and is an engineering student at the University of St. Thomas. AnnMarie Thomas is an engineering professor at the same university, and codirector of the Center for Pre-Collegiate Engineering Education.

Cup Positioning System
By Cy Tymony ■ Illustrations by Alison Kendall

YOU WILL NEED

Styrofoam cup
Pen
Staple

Paper clip
Magnet
Transparent tape

NOT EVERYONE HAS A GPS (Global Positioning System), but you can easily make a CPS (Cup Positioning System) from any cup and learn the art of orienteering!

1. Mark the cup.
Draw a crescent moon on one side of the cup. Then draw a downward-pointing arrow that rests against both tips of the moon. Write *South* at the bottom.

On the other side, draw the Big Dipper and Little Dipper constellations. These resemble pans with handles. The Big Dipper's rightmost stars point to the Little Dipper's "handle" star, which is Polaris, the North Star. Draw a downward arrow from Polaris and write *North* at the bottom.

Turn the cup over and draw the numbers 1 through 12 on its bottom in a clock formation.

2. Make a compass needle.
Straighten a staple or small paper clip. To magnetize it, rub it 30 times with a magnet in one direction only. Place the staple lengthwise on transparent tape, and fold the tape over so the staple is sealed in the middle. Now float your compass needle on a cup of water and when it comes to rest, write *N* on the north-pointing end, and *S* on the other (south-facing) end. Tape it to the side of the cup for safekeeping.

3. Find your way!
Your Sneaky CPS device is ready to help you find your directions in several ways, day or night.

Sun If the sun is visible and you know what time it is, you can easily locate directions. Turn the cup over to reveal the clock numbers. Keeping the cup level, aim the number that

represents the current hour at the sun. For example, if it's 3 p.m., aim the number 3 at the sun. Halfway between the position of the sun and the number 12 is the direction south.

Moon If there's a crescent moon in the sky, imagine an arrow pointing downward that follows the 2 tips of the moon. This points south. This trick also works fairly well with a half or gibbous (¾) moon.

Stars Position the cup so you can see the Big Dipper and the Little Dipper images on it. Now look up to find them in the night sky. It's easier to locate the Big Dipper first, then follow its rightmost stars that point to the Little Dipper's "handle," the North Star.

Illustrations by Julian Honoré/p4rse.com

Compass Add water to the cup to transform it into a compass. Float your compass needle on the surface. When it comes to rest it will point north! ◪

Cy Tymony (sneakyuses.com) is a Los Angeles–based writer and is the author of *Sneakier Uses for Everyday Things*.

Dizzy Robot

By Steve Hoefer

YOU WILL NEED

- Mini-motor, vibrator: allelectronics.com, #DCM-372*
- AG13 button cell battery
- Needlenose pliers
- 0.008" sheet tin from a hobby, art, craft, or hardware store
- Tinsnips or heavy-duty craft scissors

*If an alternative motor is needed, the requirements are:

- 1.5-to-3.5V DC
- Short (5mm, 0.2") metal tabs for power, not wire leads
- 13mm length (including shaft) x 4mm diameter (or 0.5" x 0.16") excluding the rubber casing

All dimensions are approximate

DIZZY ROBOTS ARE CUTE pocket-sized pals that dance around until they fall over. Just about anyone can build one—it only has 3 parts and requires no special skills.

1. Prepare the metal body.

The metal body holds everything together and conducts power from the bottom of the battery up to the motor.

Trace the pattern shown at right, tape it to a thin piece of tin, and cut it out using tinsnips or heavy-duty scissors. Be careful; the edges and corners will be sharp!

Bend the square base of the body at a right angle, then bend the bottom pair of wings into a rough circle to hold the battery in place. Bend the top part of the body into a circle to hold the motor. Use a pen or pencil as a rough guide to help form the shape.

2. Prepare the motor.

If the motor came with a rubberized insulating cover, remove it. Use needlenose pliers to carefully bend one of the motor's contacts around and under the motor. This will complete the circuit with the top of the battery.

3. Put it all together.

Place the battery in the base of the metal body with the negative (–) side up. Slide the motor into the upper housing and position it so the straight conductor is inside the housing and the bent conductor touches the top of the battery. Use pliers to compress the housing and hold the motor in place, being careful not to crush it.

Use It.

If everything checks out, it should already be running. Put it on a flat surface—it'll spin around and occasionally fall over. If it falls over more than occasionally, adjust the alignment of the base with pliers.

When your Dizzy Robot has had its fun, slide a small scrap of card between the top of the battery and the motor contact to turn the robot off.

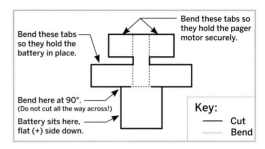

Bend these tabs so they hold the battery in place.

Bend these tabs so they hold the pager motor securely.

Bend here at 90°.
(Do not cut all the way across!)

Battery sits here, flat (+) side down.

Key:
—— Cut
.... Bend

Photography and diagram by Steve Hoefer

Steve Hoefer makes things, solves problems, and is the main brain behind grathio.com.

Sensor SenseAbility

By Cy Tymony ■ Illustrations by Mark Frauenfelder

YOU WILL NEED

RC car
Watch batteries
Buzzer
Wire

LEDs
Seltzer tablet
Universal AC adapter

WIRELESS REMOTE ALARM SENSOR

Make different types of alarms and sensors with a cheap remote control car.

Inexpensive radio-controlled cars have many sneaky adaptation possibilities. We'll use the cheap single-function type of radio-controlled toy car—the type that continuously travels forward once it's turned on until you actuate its remote control, causing it to back up and turn. You can modify the transmitter to a more compact size, and for use as an alarm trigger. You'll also modify the receiver to activate other devices, such as lights and buzzers.

CHEAP REMOTE
CONTROL CAR

1. **The transmitter is always in the Off mode until its activator button is pressed or until wires connected across the push-button leads are connected together.**

The first step is to remove the circuit board from the case. Then, substitute lightweight watch batteries for its normal battery type. For each AA or AAA battery, use one small 1.5V watch battery to supply the same voltage output.

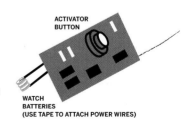

ACTIVATOR
BUTTON

WATCH
BATTERIES
(USE TAPE TO ATTACH POWER WIRES)

2. **Remove the radio receiver from the car's chassis. Unlike the transmitter, the receiver must stay On in order to operate. You can use an AC adapter (with the correct output voltage) instead.**

PCB IN CAR

The car motor is attached to the receiver with two wires. If you disconnect the motor wires, the receiver can be used for more practical purposes to indicate that a remote sensor is activated. LEDs and buzzers are perfect.

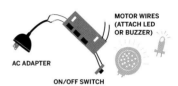

MOTOR WIRES
(ATTACH LED
OR BUZZER)

AC ADAPTER

ON/OFF SWITCH

3. **If you connect two wires across the transmitter's activator button, you can have another sensor or switch activate the transmitter to alert you of an entry breach or that your valuables are being removed, or that your basement is flooding.**

Here, two paper clips are attached to the wires. Then, a seltzer tablet is placed between the paper clips with a rubber band. If water gets near the wires, the tablet will dissolve and the rubber band will cause the paper clips to make contact and activate the transmitter.

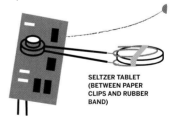

SELTZER TABLET
(BETWEEN PAPER
CLIPS AND RUBBER
BAND)

Cy Tymony (sneakyuses.com) is a Los Angeles–based writer and is the author of *Sneakier Uses for Everyday Things*.

Rear-View Power Socket

By Doug Watson ■ Illustrations by Julie West

YOU WILL NEED

12V power receptacle
   I used the Philmore TC600
   cigarette lighter plug socket
Wire, 2-conductor, insulated,
   20-gauge solid, 6" such as bell
   wire. Or 2 lengths of single-
   conductor wire

Cable ties (5) aka zip ties
Wire cutter/stripper
Soldering iron and solder
Ohmmeter or multimeter

A GPS UNIT IS a great car accessory, but the power cord hanging across the dashboard is an eyesore. Many vehicles have 12V power at the rear-view mirror, with a 2-wire plug. Tap into it to straighten up your view.

1. Wire your socket.

Unscrew the Philmore lighter socket to reveal the lugs. Strip ¼" of insulation off all ends of your wire pair. Solder one wire to the center lug (+), and the other wire to one of the outer lugs (−). Reassemble the socket.

2. Check polarity of the mirror plug wires.

Make sure your car is turned off. Use your ohmmeter to find continuity between a ground (such as the metal housing of the 12V socket in your dash) and one of the existing wires at the 12V mirror plug. The wire with zero resistance is the ground (−).

3. Connect your socket.

Push your socket wires into the mirror plug, (−) to (−) and (+) to (+), making sure they make good contact. Zip-tie them to the plug wires so they stay put. (For a permanent connection, you could solder them.) Finally, zip-tie your new 12V socket securely in its hiding place.

Your new 12V socket behind the mirror is ready to power your GPS navigation unit, radar detector, or other gadgets that are suction-cupped to the windshield.

Going Further

To mount a gadget permanently, you could solder its wires to the plug, with no socket. But I wanted my GPS to be removable. ◪

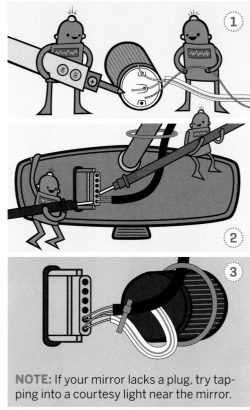

NOTE: If your mirror lacks a plug, try tapping into a courtesy light near the mirror.

Doug Watson (planithome@mindspring.com) is a fixer of all things broken; sometimes they're even improved. He recently entered his 50s, and his brother thinks he's finally playing with a full deck.

Aircraft Band Receiver

By Cy Tymony ■ Illustrations by Timmy Kucynda

YOU WILL NEED

AM/FM radio
Small Phillips screwdriver
Small flathead screwdriver

EXTENDING THE RANGE OF YOUR RADIO

The aircraft band, 108 to 138MHz, is directly above the FM band. But aircraft signals are broadcast in an AM format. Amazingly, it's possible to modify a typical AM/FM radio to receive aircraft signals in the proximity of an airport, without removing or adding any parts! Here's how.

1. Identify the radio parts.

Use a battery-powered, inexpensive, analog radio. Remove the back cover and locate the main tuning capacitor (A). It's easy to find—just turn the tuning dial, and you'll see its parts move through its clear case.

Near the main tuning capacitor you should see one or two small coils of copper wire (B) mounted on the PC board. These coils are used to limit the frequency range of the radio.

Next, locate the tuning transformers (C). They look like square, metallic boxes with tuning slots in the top. One of the tuning transformers may have a couple of small diodes near it mounted on the PC board. This is the tuning transformer that you'll adjust. Its function is to filter out AM noise.

2. Modify the radio.

Tune the radio to an FM station at the upper end of the FM band. Notice where the dial is positioned.

Spread apart the small coils near the main tuning capacitor using a small flathead screwdriver. When you finish, tune the dial and you'll notice that the broadcast stations have moved down the dial. The radio is now able to receive stations well above 108MHz.

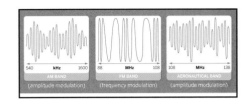

Tune the radio between stations so you can hear a slight hiss. Notice the position of the slot on the top of the tuning transformer that is near the main tuning capacitor (nearest to the small diodes). Slowly turn its screw until the hiss sound is at its maximum level. Note exactly how many turns and in which direction you turn the screw, for easy repositioning later. The radio is now able to receive AM signals in its newly expanded FM band.

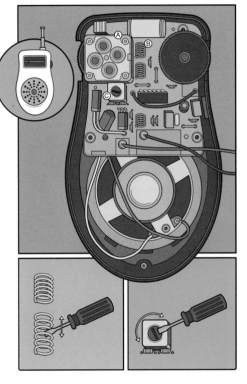

3. Test the modified radio.

Replace the radio's cover. Take it—and the screwdrivers—near an airport during a peak air traffic period. Turn the radio on the FM band with the volume up and slowly adjust the dial. You should be able to hear air-to-tower transmissions. If needed, make adjustments to the tuning coils and the tuning transformer. ◪

Cy Tymony (sneakyuses.com) is a Los Angeles–based writer and is the author of *Sneakier Uses for Everyday Things*.

Smartphone Signal Generator
By Jacob Beningo ■ Illustrations by Julie West

A SIGNAL GENERATOR IS handy to have around the lab. It's perfect for testing inputs on a new hardware design; use it with an oscilloscope to verify the behavior of your circuit. Turn a smartphone into one for less than $15!

1. Wire the adapter.

Unscrew the cap from the 3.5mm plug. Strip ¼" of insulation off the ends of each wire. Solder one wire to the plug's large tab first—this will be the ground. Then solder each of the other wires to the smaller channel tabs—these will be the signal wires. Replace the cap. Either solder or screw the opposite end of the ground wire to the black alligator clip. Then connect the 2 red alligator clips to the signal wires. Make sure to slide the alligator clip insulator down the wire beforehand so they can be replaced after soldering.

2. Download signal generator app.

Download the appropriate signal generator app for your device. On iOS, Sig Gen is one option. On Android, Waveform Generator is a great app to use. These apps will generate various waveforms at different frequencies and amplitudes. They range in price from free to a couple dollars.

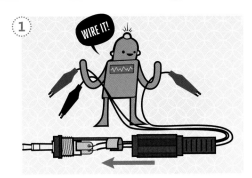

3. Connect adapter to smartphone.

Plug the adapter into the phone. Connect the clips to your oscilloscope, ground to ground and then signal to input. Turn headphone volume to maximum. Start the application. Select the waveform, amplitude, and frequency and start generating waves!

Going Further

To protect the headphone jack, a simple buffer can be added between the adapter and the circuit. For an example, and suggestions on how to use your signal generator, visit makezine.com/projects/ make-36-boards/ smartphone-signal-generator/.

Jacob Beningo (jacob@beningo.com) is a lecturer and consultant on embedded system design. He's an avid tweeter, a tip and trick guru, a homebrew connoisseur, and a fan of pineapple. What!

Shaker Flashlight Modification

By Cy Tymony ■ Illustrations by Timmy Kucynda

YOU WILL NEED

Shaker flashlight
Wire
Small Phillips screwdriver
Soldering iron
Solder

A SHAKER FLASHLIGHT has a bright LED, a NiCad battery, and an electrical generator made from a coil of wire and a cylindrical magnet. When you shake the flashlight, the magnet slides within the coil to charge the battery. Shaker flashlights provide about 2 to 4.5 volts but produce a very low output current, so there will be limits on their ability to supply power to devices. Recently Walgreens stores offered one for just $5.

1. Modify the flashlight.

Figure 1 shows the basic parts of the flashlight—the case, coil, magnet, battery, and printed circuit board. Unscrew the lens assembly and remove the screws that secure the PC board to the case. Remove an LED lead and solder 3 wires in place (Figure 2), so you can still use the flashlight normally when you're through powering other devices.

To power a device, connect the 2 wires attached to the flashlight's PC board, with the correct polarity, to the device's battery clips. When you need to use the flashlight normally, connect the LED wire and one of the PC board wires together (use color-coded wire or labels for easy identification).

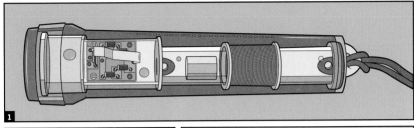

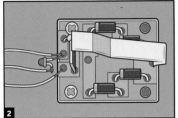
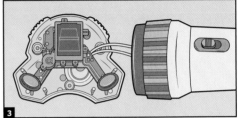

2. Adapt a device to connect to flashlight.

Select a device that requires 1.5 to 3 volts at low current, like a small electronic LCD video game. (I'm an avid collector of the free handheld games that fast food outlets supply with kids' meals. They require very little power if you turn off their sound option.) Wrap the bare ends of 2 wires tightly around the clips. Figure 3 shows the electronic game connected to your shaker power supply.

3. Go further.

A larger NiCad battery can be substituted for more capacity. You should be able to power other low-current devices such as a small Walkman-type AM radio, travel clock, or micro radio-control car. You might also (possibly) use the modified flashlight as an emergency cellphone power supply. ☑

Cy Tymony (sneakyuses.com) is a Los Angeles–based writer and is the author of *Sneakier Uses for Everyday Things*.

LED Throwies

By Graffiti Research Lab ■ Illustrations by Kirk von Rohr

YOU WILL NEED

10mm diffused LED, any color(s) 20 cents each from HB Electronic Components (LEDZ.com)

1" strapping tape one roll will make many throwies

CR2032 3V lithium batteries 25 cents each from CheapBatteries.com

½" x 1" NdFeB disc magnet, Ni-Cu-Ni plated 25 count for $13 from Amazing Magnets (amazingmagnets.com)

Conductive epoxy (optional) Weather-resistant alternative to tape. Available from Newark InOne (newark.com)

MAKE AND TOSS a bunch of these inexpensive little lights to add color to any ferromagnetic surface in your neighborhood.

1. Test the LED.

Pinch the LED's leads to the battery terminals, with the longer lead (the anode) touching the positive terminal (+) of the battery, and the shorter lead (the cathode) touching negative (−). Confirm that the LED lights up.

2. Tape the LED to the battery.

Tape the LED leads to the battery by cutting off a 7" piece of strapping tape and wrapping it once around both sides of the battery. Keep the tape very tight as you wrap. The LED should not flicker.

3. Tape the magnet to the battery.

Place the magnet on the positive terminal of the battery, and continue to wrap the tape tightly until it's all done. The magnet should hold firmly to the battery. That's it—you're ready to throw (or make a few dozen more). Throw it up high and in quantity to impress your friends and city officials. ▨

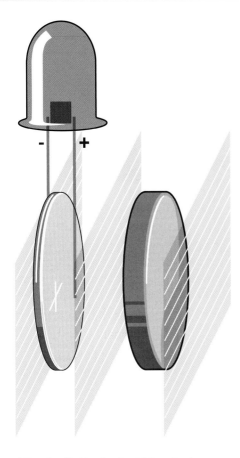

NOTE: The battery's positive contact surface extends around the sides of the battery. Don't let the LED's cathode touch the positive terminal, or you'll short the circuit.

Throwies naturally chain together in your pocket, making multi-segmented throwie bugs, which will also stick to metal surfaces if they aren't too long.

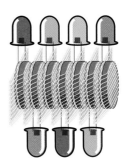

A throwie will shine for about 1-2 weeks, depending on the weather and the LED color. To get one off a ferro-magnetic surface, don't pull it, or it may come apart. Instead, apply a lateral force to the magnet base, and slide it off the surface while lifting it with a fingernail or tool.

Graffiti Research Lab (graffitiresearchlab.com) is dedicated to outfitting graffiti artists with open source technologies for urban communication.

Cup Speaker and Microphone
By Cy Tymony ■ Illustrations by Mark Frauenfelder

YOU WILL NEED

Cup
Magnet
Thin insulated wire

⅛" plug cable
Radio or music player
(with earphone jack)

TURN A CUP INTO a speaker and a microphone. People seldom think about the common devices they use everyday and even less about adapting them for other purposes. Anyone can learn real-life MacGyverisms using everyday items—you just have to be a little sneaky.

1. Turn a cup into a speaker.

A typical speaker consists of a coil of wire attached to a paper cone with a magnet mounted close by. When an audio signal travels through the wire, it creates a magnetic field. Since magnets attract and repel each other, the speaker's magnet causes the coil to push and pull the paper cone. This rapid motion vibrates the air to create sound.

This project illustrates how to use an ordinary paper or styrofoam cup, wire, and a magnet to create a Sneaky Speaker. You can use any thin wire available (e.g., from an old telephone cord) and a 1" plug cable from an old headphone cable. Use a strong magnet for this project (not the weak, ceramic type used for refrigerator magnets). A rare earth magnet, if available, works nicely.

First, wrap 10 or more turns of thin wire around a thick pen and use tape to keep it into a coil shape. Mount the coil on the back of the cup and affix it with tape.

Next, connect the coil wires to the 1" plug cable. Insert the plug into the radio's earphone jack, and turn the volume to maximum.

Hold the magnet near the end of the coil. You should be able to hear sounds emanating from the cup. If not, reposition the magnet on the back of the wire coil. Once you've located the area that provides maximum volume, tape the magnet to the back of the cup.

2. Turn a cup into a microphone.

Just as the audio signal in the coil near the magnet can vibrate the cup to produce sound, you can reverse the effect and create a Sneaky Microphone.

This project uses the same parts and setup as the Sneaky Speaker but substitutes a tape recorder for the radio.

Simply insert the 1" plug cable into the tape recorder's microphone jack. With a blank tape in the recorder, press the RECORD button. Speak loudly into the cup to record a message. Rewind and play back the tape, and you'll hear your message recorded with your Sneaky Microphone.

How it works: By speaking loudly into the cup, you vibrate the coil near the magnet. An electrical audio signal is produced in the wire that corresponds to your voice. The signal is detected and amplified by the tape recorder, which records it on tape.

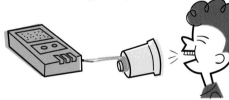

Going Further

Take apart an old speaker and study its design closely. Experiment with longer coil lengths.

Test the project with larger magnets to increase the sound level. Place the speaker in the wire coil and see what occurs. Suspend the cup with a clothes hanger frame and adhesive tape. For added volume, set up a speaker array connected in parallel to the 1" plug cable. Place the wire coil against the back of your ear and bring the magnet near the coil. ◾

Cy Tymony (sneakyuses.com) is a Los Angeles–based writer and is the author of *Sneakier Uses for Everyday Things*.

Paper Clip Record Player

By Phil Bowie ■ Illustrations by Damien Scogin

YOU WILL NEED

Paper clip, standard #1 size
Polystyrene foam packing peanut, rigid
Sandpaper, fine-grit

GOT A TURNTABLE THAT STILL TURNS? Got an old vinyl record? Make this pocket player in a few minutes for less than a penny.

1. Prepare the paper clip.
Unbend the paper clip and reshape it into a hook. Using fine-grit sandpaper, sand the long end to a fairly even, slightly dull point.

2. Join the paper clip and foam peanut.
Insert the hook-shaped clip into the peanut as shown.

3. Play a record.
Place an old 45rpm or 33rpm LP record (one you don't care about scratching) onto a turntable, and turn it on. Without pressing down too hard, carefully hold the pointed end of the paper clip in a groove as shown. You'll hear the music reproduced in remarkable fidelity, as the closed cells in the foam peanut act like a cluster of tiny amplifying speakers.

Going Further
For mellower sound, use a softer (but still closed-cell) foam peanut. For more volume, add another peanut alongside the first one, like a peanut kabob, and apply a hair more playing pressure. For more sophistication, as when playing classical selections, use an actual steel record needle with the foam peanut, available dirt-cheap online.

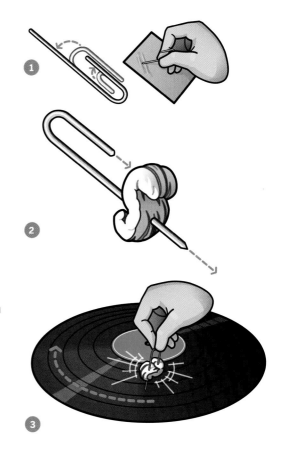

To experiment further, try straightening the paper clip and taping it firmly to the side of a small polystyrene drinking cup with a thin, translucent, plastic top snapped onto it (a half-pint deli-style container works well, too). It makes a remarkably loud speaker.

Don't have a turntable? Turn a lazy Susan into a makeshift turntable: oil the ball bearing in the base, place a piece of rubber nonskid shelf liner on the top surface, and center a 45 record on top. You have to spin it fairly fast to get enough rpm, but it does work! ◪

NOTE: It's important to hold the player by the paper clip wire to get the best sound. If you hold the peanut, your fingers tend to dampen the sound.

Phil Bowie is a lifelong freelancer with 300 magazine articles published. He has three acclaimed suspense novels and a short story collection available on Amazon. Visit him at www.philbowie.com.

Piezo Contact Mic

By Justin Emerson ■ Illustrations by Julie West

YOU WILL NEED

Piezo disc
Audio cable such as a ¼" guitar
  cable
Low-powered practice amp
Heat-shrink tubing wide enough
  to fit over the piezo

Gaffer's tape
Soldering iron and solder
Wire cutter/stripper
Screwdriver small
Heat gun or hair dryer
Hot glue gun and glue

A CONTACT MICROPHONE IS A SMALL DEVICE that can be used to amplify acoustic instruments. You don't sing or talk into a contact mic. As the name implies, it makes contact with a solid object and turns mechanical vibrations into electricity. Because a contact mic doesn't pick up ambient sounds in the room, it focuses in on one instrument without interference or feedback.

1. Aquire parts.

Gather old toys from the attic or the thrift store and pull out the blippity bleepy ones. Open them up to see if they have a piezoelectric disc or a normal speaker. Hopefully you'll find some piezos. You will need to break apart some plastic bits to get at them—just be careful not to damage the discs.

2. Wire it up.

Cut the audio cable in half. Now you have enough cable and connectors for two contact mics! Strip off a bit of outer insulation from the cut end of the audio cable. Strip and tin the signal and ground wires. Desolder the two piezo wires where they connect to the piezo. Now solder the signal and ground wires from your audio cable to those same points on the piezo. Dab some hot glue on the back of the piezo, for strength. Test it by plugging the cable into a

low-powered amp and tapping on the piezo. You should hear the sound of the amplified tapping.

3. Finish it up.

Use heat-shrink tubing, electrical tape, Plasti Dip, or epoxy to insulate the wires, the solder connections, and the piezo itself. Use gaffer's tape to attach the mic to different objects like cardboard boxes and paper cups, and acoustic instruments like guitars and kalimbas. You may get a feedback loop if the mic is too close to the amp; not necessarily a bad thing if you're a noise artist. Effects pedals can be used to flavor the sound to your taste. ◪

Justin Emerson makes experimental electronic music with hand-built and modified instruments. His band, Burnkit2600 (burnkit2600.com), performs and gives workshops and presentations on the topics of circuit bending, chiptune, and DIY electronics.

$5 Smartphone Projector

By Photojojo/Danny Osterweil ■ Illustrations by Julie West

SLIDE PROJECTORS ARE GREAT, BUT OUTDATED, and digital projectors cost a bundle. Fortunately, you can show off your mobile photos and your phone hack savvy by turning your smartphone into an inexpensive projector.

1. Prepare the projector box.

If the inside of your shoe box is a bright color, paint it black or tape up some black paper for best image quality. » On a short side of the box, trace the outer edge of your magnifying glass and cut it out. » For extended use, cut a small hole at the back of your box for your phone's power cord. Tape the magnifying lens securely in place, and make sure there are no holes to let light in.

2. Make a phone stand and flip your screen.

Bend your paperclip into a cellphone stand. » When light passes though the lens, it gets flipped, so the picture from your projector will come out upside-down. For the iPhone go to Settings > General > Accessibility and turn on Assistive Touch. Once activated, a little white orb will pop open that you can drag around the screen. Click on the orb and go to Device > Rotate

---

Photojojo is on a mission to make photography more fun for everyone! It publishes an insanely great newsletter and carries only the most awesome photo gifts and gear for photographers. photojojo.com

Screen. This will allow you to flip applications like the Photos app, which would normally rotate itself right side-up. For Android phones open the Settings menu, then select Display under the Device heading, then select "Screen rotation" and turn off the screen rotation switch to disable screen rotation.

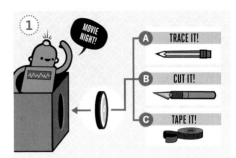

3. Find your focus.

Project onto a bare white wall or another flat, white surface. » Position your phone in its stand near the back of the box and walk the box forward or backward until the image starts to come into focus. Fine-tune the focus by moving the phone forward or backward in the box. » Set your phone's photo app to slideshow mode for a hands-free experience. » If desired, put the power cord through the hole in the back of the box and seal with a bit of tape. » For best viewing, turn the screen brightness of your phone all the way up, put on the box top, cover any windows, and turn the room lights down.

Thanks to Instructables user MattBothell for inspiring this project! ◢

Hypsometer

By Cy Tymony ■ Illustrations by Julian Honoré/p4rse.com

YOU WILL NEED

Drinking straw
Calculator
Pen
Tape
Large paper clip

Scissors and hole punch
6"×8" piece of cardboard or a
    thick plastic bag
Ruler with inch and
    centimeter markings

WANT TO KNOW THE height of a person or building? Using simple trigonometric principles, you can closely estimate the height of objects with an easy-to-make hypsometer (*hyps* means height in Greek).

1. Make it.

Tape the straw along the top of the card. Punch a small hole in the right side of the card, 10cm from the bottom. Straighten the paper clip, bend one end into a small hook, and hang it in the hole so it swings freely. Next, write numbers 0 to 12, 1cm apart, from right to left across the bottom of the card, starting with 0 directly beneath the hanging paper clip, proceeding leftward to 12. Here's a tip: If you use a thick plastic bag instead of the card, your hypsometer can be rolled up and carried in your pocket.

2. Test it.

Standing 10' away, look through the straw at the top of a friend's head. When you tilt your head (and the hypsometer), the paper clip will move leftward at an angle as gravity keeps it pointing down at the ground. In Figure 2, it points to number 1. By measuring your distance from an object (in centimeters) and the angle of tilt indicated by your hypsometer, you can calculate your friend's height.

3. Calculate it.

You'll need 2 more numbers for your calculation: the height of the card (10cm), and the height of your eye line, measured

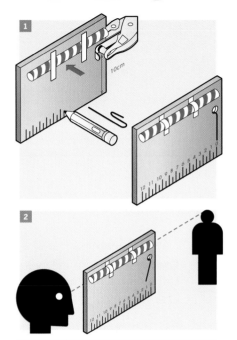

from the ground. You can also measure from the top of your head, then subtract this number from your height. For instance, if you're 5'8" tall and your eyes are 4" from the top of your head, your eye line is 5'4", or 164cm, off the ground.

Metric measurements are much easier to calculate with a hypsometer, because metric is a base 10 system. Use a metric ruler, or write equivalents on your hypsometer as shown in Figure 3.

Multiply the distance to your friend (10' or 305cm) by the angle number indicated by the paper clip (1), then divide by the card height (10); in this case, you get 30.5. Add this to your eye line height (164cm), for a total of 194.5cm. That's about 6'3" tall.

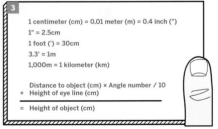

1 centimeter (cm) = 0.01 meter (m) = 0.4 inch (")
1" = 2.5cm
1 foot (') = 30cm
3.3' = 1m
1,000m = 1 kilometer (km)

$$\frac{\text{Distance to object (cm)} \times \text{Angle number} / 10}{+ \quad \text{Height of eye line (cm)}}$$

= Height of object (cm)

Now test your hypsometer with objects of unknown height, such as doors, walls, and buildings, to calibrate it and sharpen your measuring skills. ◪

Cy Tymony (sneakyuses.com) is a Los Angeles–based writer and is the author of *Sneakier Uses for Everyday Things*.

Curie Engine

By John Iovine ■ Illustrations by Dustin Hostetler

YOU WILL NEED

A bit of low-Curie nickel alloy
1" wire, or larger neodymium
    "super magnet"
Copper wire
Birthday candle

Brass screw and nut
Small steel plate
Glue or wood screws
4"×¾" wooden slat
2"×1¼"×½" wood block

Nickel wire and magnet, or a full kit, are available for purchase at makezine.com/go/heatengine

CHANGING THE TEMPERATURE OF NICKEL WIRE turns heat into motion.

1. Put it together.

Glue or screw the steel plate and 4" slat to the wood block base as shown below. Drill a pilot hole, and screw the brass screw at the top of the slat. Drill a ¼" hole in the base about ⁷⁄₁₆" from the steel plate end. For complete measurements, go to the website listed above.

Wrap about 1" of nickel alloy wire around a pencil to make a coil. Take 4" of copper wire, and twist the coil onto one end. Make a 90° bend in the copper wire 1½" down from the end with the coil, and cut the other end so there's 1" after the bend. Hook the wire bend onto the brass screw, with the coil on the side nearer the steel plate.

2. Test the wire and magnet.

Place the neodymium magnet onto the metal plate. The wire should swing up to meet the magnet, pulled by the nickel alloy coil. Push down and release the wire, and it should swing back up. Insert a birthday candle in the base, and adjust the magnet position and wire so the coil is suspended above the wick.

3. Light the candle.

The wire should swing down and up repeatedly, out of and back into the candle flame. Like ordinary iron or steel, the nickel-iron alloy is magnetic, but it has a much lower Curie point—the temperature above which it loses its magnetic properties. (The copper wire and brass screw are not magnetic.) When the coil heats up to this point, it is no longer attracted to the magnet, so it falls away and out of the flame. As it cools, it becomes magnetic again and swings back up, over and over again, until the flame burns too low to touch the coil.

John Iovine is a science and electronics tinkerer. He has published books and articles, and he owns and operates Images SI Inc. (imagesco.com). To learn more about this topic visit: imagesco.com/articles/heatengine/HeatEngine.html

Sound Sucker
By William Gurstelle ■ Illustrations by Damien Scogin

USING THIS "SOUND SUCKER" DEVICE allows you to experience a curious sensation: it's as if sound is not only being blocked, but actually sucked away from your ear.

The sound sucker works on a narrow range of frequencies. My testing showed it most effective a few cycles to either side of 660Hz (depending on the amount of gelatin), and the effect is most noticeable in a room with a wide spectrum of ambient noise frequencies.

Can you explain this acoustic phenomenon? Or better yet, can you draw a simple diagram showing how you think it works?

1. Start cookin'.

Prepare the gelatin according to box directions. Before it sets, reserve about a ¼ cup of the liquid for the project.

2. Pack your mug.

Pour the ¼ cup of gelatin into the mug. Place coffee stirrers in the mug, packing it as densely as possible. When done, it looks something like the compound eye of an insect.

Make: contributing editor William Gurstelle is the author of the "Remaking History" column as well as popular DIY books including *Backyard Ballistics* and *The Practical Pyromaniac.*

3. Let it set.

Place the stirrer-packed mug in the refrigerator until the gelatin sets. The idea is that the gelatin seals the bottom tip of each stirrer.

Suck Up the Sound

In a place with medium-to-loud ambient noise, hold the mug next to your ear, tilted sideways so the opening is facing your ear. If there's enough sound in the 660Hz range, you'll notice a sudden drop-off in acoustic energy (noise) when you bring the sound sucker near.

If you have a frequency generator, test a range of frequencies. Besides the 660Hz tone, it also attenuates some higher frequencies. ◪

Sneaky UV Ink Password Protector

By Cy Tymony ■ Illustrations by Damien Scogin

YOU WILL NEED

Invisible ink (UV) marking pen
UV LED
Paper

Option 1:
Wide barrel pen
Coin cell battery
Tape

Option 2:
LED pocket light
Screwdriver

Option 3:
Spare remote control

WITH SO MANY COMPUTER passwords to remember these days, it's necessary to safely store them somewhere—but not in a computer file. So where?

Write them with an ultraviolet (UV) invisible ink pen so the information can't be seen by the naked eye. Only shining a special UV light on the passwords will reveal them. Use paper with written material already on it, so people think the visible words are all that's there.

But how do you prevent someone else from using your UV light? By hiding it too! Here are 3 ways to render your UV light unfindable to the casual snooper.

1. Hide a coin cell battery and UV LED.

Get a UV LED (available online for under $1) and a small watch battery (you could remove them from a standard UV LED light set). Wrap them together with tape, and hide them inside a wide-barrel pen. If needed, you can retrieve the parts, squeeze the battery between the LED's legs, and check your passwords.

2. Install a UV LED in an LED pocket light.

Use a screwdriver to open an LED pocket light. Remove the standard LED and replace it with a UV LED. Now shine the invisible beam at your message to reveal it. This method works well, but someone might be suspicious that you keep a "nonworking" LED pocket light around.

3. Hide your UV LED in a remote control.

To be really sneaky, substitute a UV LED for the infrared LED found in a TV or DVD remote control. Use a discarded remote, or buy a universal remote from a dollar store. If people try to use the remote, they probably won't detect anything unusual, because a remote doesn't make a bright light anyway! ▨

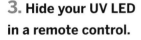

Cy Tymony (sneakyuses.com) is a Los Angeles–based writer and is the author of *Sneakier Uses for Everyday Things*.

Part IV: Home and Outdoors

EASY 1+2+3 PROJECTS CAN BE PRACTICAL AND USEFUL, as well as fun. The projects that follow include tips and tricks for the kitchen, useful ideas for furniture and décor, and suggestions for recycling and reusing things that might otherwise end up as landfill.

Corky Mork is a lifelong Maker whose many interests range from LED projects with microcontrollers to backyard railroading. In "Stud Chair," he outlines how to make a sturdy chair out of a single 2x4 board.

Regular *Make:* magazine readers will know that one of the favorite building materials of Makers everywhere is good-old PVC pipe. After all, it's versatile, strong, and economical! The standard white pipe gets a little boring though. That's why Sean Michael Ragan demonstrates a method for creating a little variety in "Stain PVC Any Color."

Our prolific "DIY Hacks and How Tos" author Jason Poel Smith contributes a quintet of clever projects. The "Clothes Folding Board" will help you put away your freshly laundered shirts quickly and efficiently. His "Custom Soda Cooler" is the perfect summer picnic companion. Trying to spray hard-to-reach areas can be frustrating, but Jason's "Omnidirectional Spray Bottle" hack solves the problem. "Keyboard Refrigerator Magnets" suggests a way to reuse old computer equip-ment that may give you more ideas along similar lines. And his "Realistic Duct Tape Rose" demonstrates once again the unending versatility of duct tape with the perfect handyman's valentine.

In "Box Fan Beef Jerky," *Make:* magazine engineering intern Paloma Fautley passes along a tasty technique for turning a box fan and some air filters into a food dehydrator. YouTube's CoolGadgetGuy Tom Fox has a nifty baking tip that he shares in "Burnt Cookie Deflector." You'll never again make cookies without it!

Design engineer Gus Dassios returns (see "Dice Popper" earlier in this book) with two useful ideas. "Paper Clip Paper Holder" presents a handy desktop helper made from a metal coat hanger, while "World's Cheapest Monopod" fashions a quick and easy camera stand from a broomstick.

"Skip" Arey usually favors mechanical or electronics projects relating to his two main hobbies—bicycling and amateur radio. But the "Trash Can Composter" project ties into his passion for scrounging and repurposing. He'll show you how to turn an old trashcan into a composting canister for creating rich, fresh soil.

You'll find all these and even more handy household projects in the pages that follow. ◹

Stud Chair
By Corky Mork

RECENTLY, AT MY LOCAL HOME CENTER I stopped to look at the 2×4 studs. They cost just a few dollars, and I often take a few home "just to have." They're handy for workbench legs, sawhorses, shelving, mailbox posts, temporary staging, and many other uses.

I got to thinking, What could I make with just one 8' 2×4? I got out my sketchpad and worked it through. . .Yes! I could do it.

I set to work with my table saw, and in less than an hour, I had a perfectly serviceable chair! It's not the prettiest or most comfortable, but for the price in materials and effort, it's hard to beat.

1. Cut the 2×4.

Cut the 2×4 into 1 32" and 4 16" lengths, then rip these pieces lengthwise. The following rip measurements assume a ⅛" kerf.

Rip the 32" piece into thirds (1³⁄₃₂" thick) for the legs, then cut one of these into 2 16" legs. Rip the 16" pieces into 8 slats ¹¹⁄₁₆" thick.

2. Screw and glue.

A square can help keep things, well, square during assembly. For each joint, square it up, drill pilot holes to avoid splitting, and assemble with glue and 2 drywall screws. The glue will

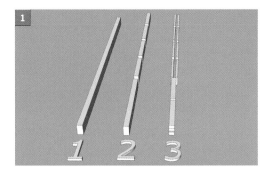

help keep the chair from racking (twisting out of square), especially on the leg joints.

First build the 2 side frames; one is mirrored from the other. Add slats to the front and the seat back, then add the seat slats.

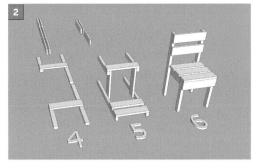

3. Finish it, or not.

You can paint, stain, and decorate your chair any way you like. I kept my first one au naturel and left it outdoors to develop a rustic look.

What can *you* make with an 8' 2×4? Maybe a better chair? A table? Or what? ◪

Illustration and photo by Corky Mork

Corky Mork (tiny.cc/corkysprojects) has been a Maker since he was kid. In addition to woodworking, he enjoys working with microcontrollers, stage effects, toys, and outdoor projects, including a kid-powered backyard railroad.

Stain PVC Any Color
By Sean Michael Ragan ■ Illustrations by Damien Scogin

> **YOU WILL NEED**
>
> Nitrile gloves
> Safety goggles
> PVC cleaner Check the label and make sure it contains tetrahydrofuran. I used Oatey Clear Cleaner, a product used to prepare PVC pipe for gluing.
> Volumetric pipette, measuring 1mL
>
> Solvent dye I found Rekhaoil Red HF (Solvent Red 164), Rekhaoil Yellow HF (Solvent Yellow 126), and Rekhaoil Blue (Solvent Blue 98) on eBay by searching "petroleum dye."
> Paper towels
> Bent wire hanger (optional)

PVC PIPE IS GREAT, but it's kinda ugly—it only comes in white, gray, sometimes black, and clear. Sure, you can paint it, but paint can flake and can screw up dimensional tolerances. Stain doesn't flake or add thickness, so the pieces will still fit together.

1. Mix the stain.

Visit makezine.com/projects/make-30/stain-pvc-any-color-you-like/ for a list of dye-to-PVC-cleaner ratios for red, orange, yellow, green, blue, indigo, violet, brown, and black.

Using your pipette, draw up the required volume of each dye and transfer it to the PVC cleaner container. Be careful not to cross-contaminate the dyes. Note that solvent dyes are very strong; 1 ounce goes a long way.

Close the can lid tightly. Wipe off any stray liquid on the outside of the can. Gently shake for about 15 seconds to mix.

 **⚠ caution:** Work in a well-ventilated workspace and wear gloves and goggles when handling the solvent or dye.

2. Apply the stain.

You can use a holder for the PVC, such as a piece of bent wire hanger. Generously slather the stain onto the pipe using the cleaner can's built-in applicator. Work quickly, smoothing out streaks before they have time to dry.

3. Dry and test.

The solvent will dry quickly—an hour will be more than enough. Once dry, the stained PVC should be able to pass a "white glove test" and not transfer even a small amount of color to any-thing that touches it.

NOTE: Dyes can fade over time; try using light-fast dyes or adding UV stabilizers. ◪

Sean Michael Ragan is descended from 5,000 generations of tool-using hominids. Also he went to college and stuff.

Clothes Folding Board

By Jason Poel Smith ▪ Illustrations by Andrew J. Nilsen

FOLDING CLOTHES IS A boring and laborious chore. So I tried to figure out a way to speed up the process using materials that I had lying around. My solution was to make a folding board out of cardboard and duct tape, like the ones used to fold shirts at retail stores.

1. Cut out the panels.

To make a folding board, you'll need 6 cardboard panels. Each panel should have the dimensions of a folded shirt.

So fold one shirt by hand and measure its dimensions. Then cut out 6 pieces of cardboard this size.

2. Assemble the board.

Lay the cardboard panels on the floor in 2 rows of 3. Space them out so that there's a ¼" gap between each piece. Next, tape the 3 panels in the top row together.

Then individually tape each panel in the top row to the panel below it.

Turn the whole assembly over and apply tape to the back of all these same gaps.

Press the 2 sides of tape together to seal them.

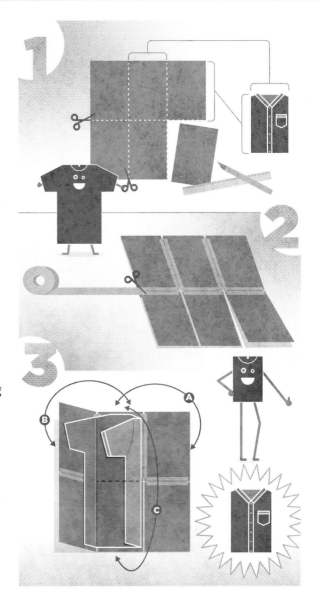

3. Use the folding board.

Place the folding board on your work surface. Then lay a shirt face down on top of it.

Fold one side panel over and back.

Then fold the other side over and back.

Lastly, fold up the bottom center panel. You should now have a perfectly folded shirt. This tool can drastically speed up the folding process, and every shirt will be uniformly folded.

See the how-to video and more photos at makezine.com/go/diy-clothes-folding-board. ⤢

Jason Poel Smith makes the "DIY Hacks and How Tos" project video series on *Make:*. He is a lifelong student of all forms of making, and his projects range from electronics to crafts and everything in between.

Give Old Work Jeans New Legs
By Gregory Hayes ■ Illustrations by Julie West

YOU WILL NEED

Pair of jeans
Fabric scissors
Straight pins
Sewing machine
Upholstery thread

DON'T LET A TORN seam doom your veteran work jeans to the landfill. With just a few stitches, they can return to the workshop with honor as a handsome, durable tool wrap for wrenches, chisels, and the like.

1. Cut 'em.
Starting at the pant cuff, cut up the length of the inseam, staying quite close to the tucked edge of the seam. When you reach the crotch seam, turn 90° and cut around the leg; take care not to cut through any pockets in case you decide to repurpose more than just the legs.

2. Fit 'em.
With one flap of the leg material open, lay out your tools evenly with elbowroom for easy rolling, then peg between them with straight pins. If the leg's too long for all your tools, trim the excess (or get more tools).

3. Stitch 'em.
Remove the tools, then straight stitch on your sewing machine in the places you pinned, using upholstery thread. To get the top flap to always fold the same way, stitch along the folded

top edge. If the pockets are too deep for some tools, add a stitch to shallow those pockets. For added finish and durability, fold over and stitch all edges; for a rough look, don't stitch anything you don't need to.

Going Further

Trim out the waistband (including zipper) to create a buttoning strap to secure the roll around its top. For more ideas visit makezine.com/projects/make-34/123-give-old-work-jeans-new-legs/. ⬈

Gregory Hayes wrangles photos for *Make:* and knows lots of other ways to destroy good jeans.

Custom Soda Cooler

By Jason Poel Smith ■ Illustrations by Andrew J. Nilsen

YOU WILL NEED

Vinyl sheet I used a vinyl tablecloth.
Flexible foam sheet such as craft foam

Sewing machine or spray glue
Velcro or a zipper

SUMMER IS HOT. So why not make a custom cooler to keep your drinks cold! Here's how to make a flexible cooler that perfectly fits a 12-pack of soda. Take the whole pack out of the fridge, slip it into your custom cooler, and go.

1. Make the sides.

Start by determining the shape of your cooler. Trace the outline of each side onto the foam. Then cut out each outline.

Apply a generous amount of spray glue to one sheet of vinyl. Then stick on the foam cutouts, spacing them out by at least 1".

Apply spray glue to a second sheet of vinyl and stick that on top.

Dry overnight, then cut out.

2. Assemble the cooler body.

Lay out the sections as they will be assembled, and glue or sew the pieces together.

After attaching the sides together, trim the seams to no more than ½".

3. Invert and add a closure.

Carefully turn your cooler inside out, so that all the seams are on the inside. If you're worried about breaking a weak seam, you can secure your seams with binder clips before inverting the cooler.

Finally, add a zipper or velcro that will hold the opening of the cooler closed. Attach this by gluing or sewing it in place. You're ready to chill.

See step-by-step photos and video and share your cooler designs at makezine.com/go/custom-cooler. ◢

Jason Poel Smith makes the "DIY Hacks and How Tos" video series on *Make:*. He is a lifelong student of all forms of making, and his projects range from electronics to crafts and everything in between.

Paper Clip Paper Holder

By Gus Dassios ■ Illustrations by Damien Scogin

YOU WILL NEED

Wire clothes hanger
  or other stiff wire
Paint and primer (optional)
Wire cutters or hacksaw
Lineman's pliers

File
Electrical tape
Bench vise or another way of
  making a clean bend
Paper clip (optional)

WHAT'S BETTER AT HOLDING paper than a paper clip? A bigger paper clip! Here's a very simple project that was completed in about 5 minutes. Finding the parts and tools took longer!

1. Cut the wire.

Cut a 15" length of stiff wire. Wire cutters make the cut easy, but a small hacksaw will work, too.

You can use the bottom of a wire clothes hanger, so that you don't have to straighten out any of the bends. Naturally, a longer length will make a larger clip and a shorter length will make a smaller clip.

File the sharp ends until smooth.

2. Bend it into a paper-clip shape.

Look at a normal paper clip to help you determine where and how to make your bends.

Someone with strong hands should be able to bend the wire without any tools, but lineman's pliers make it easier to twist the wire into a large version of a traditional paper clip.

Wrap some electrical tape around the teeth of the pliers so they don't dig into the wire too much.

3. Make a 90° bend.

For the final bend— required so that the clip can be freestanding— use a vise. Secure the bottom quarter of the clip in the vise, and then bend the top over into a 90° bend.

After removing the clip from the vise, bend it by hand another 10° farther, so that it won't tip over easily.

If the coating remains intact on the clothes hanger wire, you're finished. If you chipped the coating during the bending process, or if you used bare metal wire, you can prime and paint your paper clip paper holder.

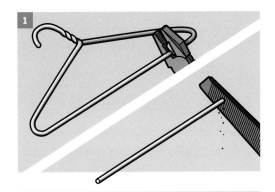

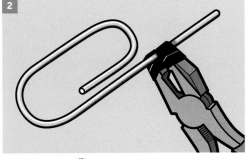

Just MAKE it

Gus Dassios lives, designs, and builds in Toronto, Ontario.

Omnidirectional Spray Bottle

By Jason Poel Smith ■ Illustrations by Julie West

YOU WILL NEED

Spray bottle, 16oz or bigger
Flexible tubing, ⅛" ID, ¼" OD, such as aquarium air line
Stainless steel nuts, ¼" (5)
Scissors
Glue (optional)

MODIFY A SPRAY BOTTLE so it works when held at any angle. If you replace the hard suction tube with flexible tubing and a weight on the end, the tubing will naturally fall to the lowest point of the container.

1. Cut the tubes to length.

Cut the original spray bottle tube, leaving about 1" sticking out past the screw cap. Then cut the flex tubing about 1" longer than the cut portion of the original tube.

2. Attach the 2 cut tubes.

The suction tube on a typical spray bottle has an outer diameter between ⅛" and ¼", so the flex tubing should make a good seal without any adhesive. Slide it onto the original tube until you get a firm seal, with at least ¼" of overlap.

If you don't get a satisfactory seal, use glue. With both tubes clean and dry, apply a thin layer of glue around the lower ½" of the original tube. Slide the flex tubing onto the original tube, overlapping by at least ½", and slowly twist the flex tube to help spread the glue evenly. Let the adhesive completely cure before continuing.

3. Add the weight.

Twist a nut onto the end of the flex tubing and turn the sprayer upside down to see how low the tube hangs. Add nuts until the tube hangs almost down to the cap. The stiffer the tubing, the more nuts you'll need. I used 5.

Use It.

Fill the bottle with liquid and enjoy spraying up, down, and upside down! If the tube gets stuck, give the bottle a gentle shake. ▨

NOTE: Even stainless steel eventually corrodes. After each use, remove the sprayer assembly to dry.

Jason Poel Smith is a helicopter tooling engineer. When he's not inventing, he's spending time with his amazing family.

Burnt Cookie Deflector

By Thomas R. Fox ▪ Illustrations by Damien Scogin

YOU WILL NEED

Aluminum foil heavy-duty
Cookie sheet roughly the same size as the
cookie sheet you use to bake the cookies on

PROBABLY THE TOUGHEST PART OF BAKING great-tasting cookies is avoiding a hard, burnt bottom! While the problem could be that the oven's thermostat is inaccurate, the cause is more likely that infrared radiation is striking the bottom of the baking sheet, which heats it several hundred degrees hotter than the oven's air temperature. This really hot baking sheet is what burns the bottoms of the cookies.

To solve the problem, keep the baking sheet as cool as possible by using a radiation deflector shield, made by covering a second baking sheet with shiny aluminum foil.

1. Cut the foil.

Cut a length of foil 8" longer than the cookie sheet (or 2 lengths if your sheet is wider than the foil).

2. Cover the baking sheet.

Center the foil over the baking sheet, shiny side up, so you have 4" of overlap on each end, and secure it in place by folding the extra foil around the bottom of the sheet.

3. Set inside oven.

Place the oven rack for the cookies in the middle position of the oven, and move the deflector rack to the lowest position.

Invert the newly made deflector so the foil side faces down and the bottom of the cookie sheet faces up, and place it on the lowest rack.

Once the cookies are ready for the oven, simply place the baking sheet on the rack above the deflector. There's no need to adjust oven temperature or baking time.

How It Works

In an electric oven, the heating element gets close to 1500° F and radiates intense infrared to the cookie sheet. In a gas oven, the burner does not get quite hot as an electric element; but, even with gas, it isn't uncommon to get burnt cookie bottoms! The shield's heat capacity (amount of heat energy needed to change a body's temperature) protects the bottom of the cookie sheet from the most intense infrared radiation. ◪

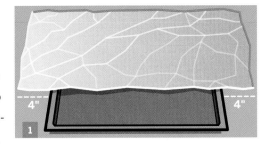

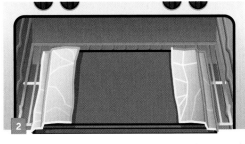

Photo by Tom Fox

Tom Fox (magiclandfarms.com) is a book author and magazine editor.

Box Fan Beef Jerky

By Paloma Fautley ■ Illustrations by Julie West

YOU WILL NEED

Beef
Marinade
Knife
Kitchen container or plastic
   bags for marinating

Air filters, nonfiberglass (3-4) for
   furnace or air conditioner
Bungee cord
Box fan

JERKY IS THE ON-THE-GO SNACK that has fueled Native Americans, pioneers, and astronauts alike. And it's easy to make at home. This method, popularized by Alton Brown, uses an ordinary box fan and air filters to dry the meat.

1. Slice it.

Buy your desired cut of beef (sirloin is a good choice), and slice it into thin strips using a long, thin blade.

2. Marinate it.

Choose a marinade and let the meat soak in the refrigerator for 6–8 hours. Make sure to include ingredients that help the meat stay moist and tender. Honey is a good choice.

3. Dry it.

Remove the meat from the marinade and pat dry with a paper towel. Place the strips on top of a large, clean air filter (make sure it contains no fiberglass). Continue stacking meat and filters until you

Paloma Fautley

run out of meat. Place one final air filter on top and put the stack of filters on a box fan. Use a bungee cord to secure the filters, then turn on the fan.

Let the fan run and enjoy that sweet, sweet meat smell for 8–12 hours. Once the meat is dehydrated, enjoy!

Check out step-by-step photos at: makezine.com/projects/box-fan-beef-jerky. ◪

Paloma Fautley is currently pursuing a degree in robotics engineering and has a wide range of interests, from baking to pyrotechnics.

Trash Can Composter

By Thomas J. Arey ■ Illustrations by Damien Scogin

I LIKE TO PICK things out of trash cans and reuse castaway items. Here, I repurpose the trash can itself to facilitate recycling organic waste into beneficial compost.

Commercial composting canisters can be costly, but they're simply a place to allow natural microbial processes to convert waste matter into a dark, fresh-smelling soil. Commercial versions allow air and some water to get in, and sometimes a way to mix.

Most home trash cans fail when part of the bottom wears away, leaving a hole. Such a trash can is perfect for this project, since we're just going to add more holes anyway.

1. Clean the trash can.

Scrub the trash can thoroughly inside and out to ensure that no inorganic waste remains. If you're squeamish about this, buy a new one.

It's helpful but not necessary to have a lid. If the original lid has gone missing, I'll leave it to you to come up with another solution.

2. Drill air holes.

Use a drill with a 1" spade bit to make air holes. Space the holes about 3" to 4" apart over all sides of the trash can. Drill plenty of holes, but don't compromise the structural integrity of the trash can. Avoid the corners to maintain the trash can's strength. Drill holes on the bottom to help with drainage.

3. Start composting.

Theories about composting are as numerous as the holes you have drilled. General rules:

▶ Keep the compost material damp, not wet.

▶ Mix brown material (such as leaves) in with green material (such as grass clippings).

▶ Use uncooked food scraps; no meat.

▶ Don't allow pet waste or anything treated with pesticides into your composter.

▶ Turn the compost material regularly. If your can has a tight-fitting lid, you can lay it on its side and roll it around on the ground. ◪

T.J. "Skip" Arey has been a freelance writer to the radio/electronics hobby world for over 25 years and is the author of *Radio Monitoring: A How To Guide*.

◪ **TIP:** Nothing goes to waste in this project! The 1" cutouts made by the spade bit can be used as insulating washers in other projects.

Keyboard Refrigerator Magnets

By Jason Poel Smith ▪ Illustrations by Julie West

YOU WILL NEED

Computer keyboard

Sheet magnets **new or recycled from magnetic advertisements, car signs, or business cards**

Screwdriver

Scissors

Wire cutters

Pliers, **needlenose**

Hot glue gun

I ALWAYS TRY TO FIND WAYS TO REUSE the parts from my old electronics. I had a couple of old keyboards lying around, so I decided to use the keys to make alphabet refrigerator magnets.

1. Remove the keyboard keys.

Use a narrow screwdriver to pry the keys off the keyboard. In most cases they will just pop right off.

Look at the backside of the keys. If the mounting tabs stick out past the body of the key, then you need to trim them so that you can mount the magnets flush with the backside. You can either cut the tabs with wire cutters or just break them off with needlenose pliers.

2. Cut the magnets into squares.

Now get some magnetic advertisements or magnetic business cards. Use scissors or a knife to cut the magnetic sheet into squares that are the same size as the back of the keys.

3. Glue the keys to the magnets.

Apply a large drop of hot glue to the back of one of the keys. Then press the key onto one of the magnetic squares that you cut out. You want the bare magnet side to be facing out so that it can stick to the fridge better.

Repeat this process with all the keys. ◪

Check out step-by-step photos at:
makezine.com/projects/keyboard-refrigerator-magnets.

Jason Poel Smith makes the "DIY Hacks and How Tos" project video series on *Make:*. He is a lifelong student of all forms of making and his projects range from electronics to crafts and everything in between.

Jason Poel Smith

World's Cheapest Monopod
By Gus Dassios ■ Illustrations by Damien Scogin

MOST PEOPLE ARE FAMILIAR WITH tripods. They have three legs and are great for setting up at a location and taking photos from that one spot. Monopods, on the other hand, have only one leg, so they can be quickly moved from place to place.

But with only one leg, monopods aren't very stable. This one has a spike at one end that can be pushed into the ground for timed photos. Or you can take fixed-height orbital photos, which is useful for projects like Print Your Head in 3D (makezine .com/projects/make-ultimate-guide-to-3d-printing/print-your-head-in-3d/) in which you take a series of photos, create a digital mesh online, and then print out your subject on a 3D printer.

1. Prepare the handle.

If it has a rounded end, cut off or sand the end so it's flat. Pre-drill each end of the broom handle: one with the 3/16" bit, the other with the bit sized for your nail.

Pre-drilling allows the hanger bolt and nail to go in easier and reduces the risk of the wood splitting.

2. Attach the hardware.

Secure the hex nut on the hanger bolt, and then attach both to the broom handle using a wrench.

On the other end, install the nail by carefully hammering it into the other pre-drilled hole. After it's secure, use a hacksaw to cut off the head of the nail. This will allow the monopod to be stuck into the ground.

3. Attach the camera.

The final step is to screw on the camera. Most cameras have a ¼"-20 threaded hole on the bottom. Without the threaded hole, it cannot be mounted to a monopod or a tripod. If you secure the monopod into the ground, make sure it's steady before letting go. ◪

Gus Dassios lives, designs, and builds in Toronto, Ontario.

Realistic Duct Tape Rose
By Jason Poel Smith

YOU WILL NEED

Duct tape, red, ~3-4' per rose
Duct tape, green, ~2-3' per rose
Floral wire or other stiff wire

Markers, red and green
Scissors
Wire cutters, optional

SURPRISE YOUR SWEETHEART WITH a realistic, hand-made rose crafted from an unexpected material—duct tape!

1. Make the petals.

Cut off a 1' length of wire and a 4" piece of red duct tape. Stick the wire to the tape so that they overlap by 1½". Then fold the tape over the end of the wire and stick the 2 sides together, leaving about 1" of sticky tape exposed.

Cut the petal to shape by rounding off the corners, and color in the cut edges with markers. Make about 10 petals.

2. Build the bud.

Take the first petal and loosely roll it into a tube. Then softly wrap each additional petal around the previous one. As you go, bend a gentle curl into each petal to give it the shape of an actual rose.

3. Make the stem.

Build the stem by twisting the wires together and wrapping them in green duct tape.

To make the sepals, take several pieces of green duct tape and fold one end over, similar to the way the petals were made. Cut out 5 small triangles with the exposed tape at the base, and color in the cut edges. Then attach these to the base of the flower. Your duct tape rose is complete. ◪

See the how-to video and more photos and tips at makezine.com/go/duct-tape-rose.

Jason Poel Smith makes the "DIY Hacks and How Tos" project video series on *Make:*. He's a lifelong student of all forms of making, from electronics to crafts and everything in between.

Jason Poel Smith

Urban Survival Button

By Cy Tymony ■ Illustrations by Tim Lillis

EMERGENCY SURVIVAL KITS ARE designed for the wilderness and enclosed in pouches or mint containers. But what about survival situations in the city?

Prepare yourself by making a Sneaky Button Survival Kit, which contains practical items you can use to protect yourself in urban situations.

1. Insert the magnet.

Get a hollow garment button from an old coat or a sewing supply store. It has 2 parts: a button and a separate cap or cover. Open the cover, put the small magnet inside, and press the button back together.

2. Attach to your jacket.

With the scissors, remove a button from your jacket and punch a small hole where it was attached.

Remove the key ring (if supplied) from the key fob. Push the fob's extendable cord through the jacket's small hole and loop it through your button's shank.

3. Attach the mirror.

Cut a small, round piece of velcro and stick it to one side of the key fob. Peel the other side of the velcro backing and press the back of the small mirror to it.

Use It.

Pull out the key fob and use the mirror to see behind you while you're walking, or when using an ATM.

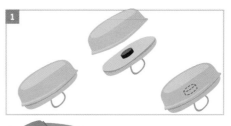

To test for counterfeit currency, bring the button near the edge of a folded bill. If the bill is legitimate, it'll move toward the magnet, due to the iron particles in the currency's ink.

Going Further

Store more emergency items inside the button:

Fire: Tear off some match heads and a small piece of the matchbook striker material. (Wrap the striker in tape so the matches don't strike accidentally.)

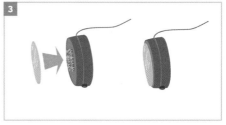

Compass: A small, straightened staple can make a handy compass in a pinch. Simply stroke it multiple times with the magnet.

Light: Store a superbright LED and a small 3V battery (or two 1.5V batteries). Insulate the batteries with tape so they won't drain their charge. ◪

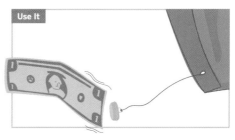

Cy Tymony (sneakyuses.com) is a Los Angeles–based writer and is the author of *Sneakiest Uses for Everyday Things*.

Easy Boombox
By Matthew T. Miller ■ Illustrations by Julian Honoré

YOU WILL NEED

Earbuds
Paper cups
Pocketknife

Audio source
Velcro strip

MANY EARBUDS CAN PLAY at high volume without distortion. Use them to drive paper cups, and the sound is surprisingly good. I first put this together when I really needed to listen to some jams at work. Put the ghetto back into the blaster!

1. Prepare the cups.
Cut small Xs in the bottoms of 2 paper cups.

2. Insert the earbuds.
Carefully insert the earbuds into the Xs. They should fit snugly.

3. Turn on the jams and rock out.
See how much amplification you can get out of paper cups!

Matthew T. Miller

Use It.

My earbuds do play at very high volumes without distortion. . .so the sound from my ghetto blaster is actually really good. This is a cheap, easy trick and will save you serious money. ◪

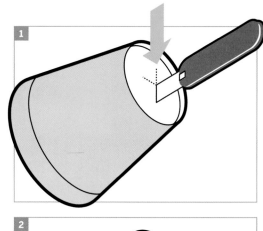

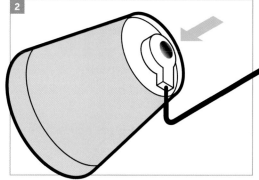

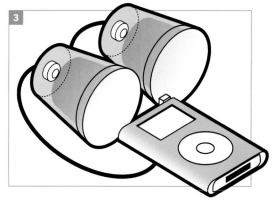

Matthew Teague Miller is a 36-year-old father who lives in San Pedro, CA. He loves to create things out of household objects.